A Citizen's Guide To Plastics In The Ocean: More Than A Litter Problem

Third Edition

Foreword by
Ted Danson

Written, compiled and edited by
Kathryn J. O'Hara, Suzanne Iudicello, and Rose Bierce

Artwork by
Jill Perry Townsend and Susie Gwen Criswell

Published by
Center for Marine Conservation
(formerly the Center for Environmental Education)
Washington, DC

Funding for this publication was provided by the Society of the Plastics Industry and the National Oceanic and Atmospheric Administration. CMC is also grateful to those who provided editorial assistance and expert review.

CONTENTS

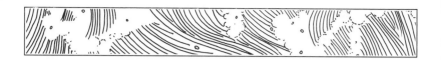

FOREWORD

A sk the average person to describe the ocean and the responses will probably be along the lines of "huge," "deep," "powerful," "teeming with life," "unpredictable," "ever-changing," or "rich." Some might even add "mysterious" and 'life-giving." But few of us would respond "finite." We just can't imagine the ocean—vastly wide, vastly deep—as having limits. We have survived and even flourished on its bounty. We have measured, examined, and explored its depths for centuries and we're still learning amazing things. And we have dumped our refuse into its limitless expanse for generations.

But we are finding out the ocean is not limitless. It can take only so much abuse, so much careless disregard. Its creatures can devise only so many ways to adapt to our activities and our artifacts. The rapid increase in the last fifty years of the production and application of plastic has brought this lesson home in ugly and disturbing ways.

Just look at the headlines—recreation beaches closed due to a dangerous level of trash on shore—commercial fishermen competing with their own lost gear—seals and seabirds strangled by the unbreakable rings of six-pack holders—sea turtles eating plastic bags, mistaking them for jellyfish—these are the results of our misuse of plastic, especially in how we dispose of it. Plastic doesn't disappear when we throw it away. We can't afford to ignore plastics and other non-degradable items just because we're finished with them. The ocean just can't tolerate it. We must learn to be responsible plastics users.

This book shows us how. Author Kathryn J. O'Hara, a biologist and Director of Marine Debris and Entanglement Programs at the Center for Marine Conservation, has written an overview of the problems that plastic debris causes in the marine environment and describes how citizens and public officials are working together to reverse the trend toward an "ocean of plastic." Written in cooperation with the Society of the Plastics Industry and the

National Oceanic and Atmospheric Administration, the book familiarizes readers with the types of plastic causing problems, where it comes from, and how it affects the ocean and marine life. Along the way, Ms. O'Hara introduces us to some special persons who have been affected by plastic marine debris, and who challenge our perspectives and our complacency.

The ultimate goal of this book is to arm citizens with information on the problem and to inspire active participation in the solutions. It will take many years (plastic is durable, remember) and will require the cooperation of citizens, industries, and federal and state officials, but Ms. O'Hara is confident, and I share her confidence, that we can overcome the obstacles. We must, if we are to maintain the ocean's health and conserve its precious living resources.

Ted Danson
Actor, President of
American Oceans Campaign

INTRODUCTION

The oceans—a domain of almost incomprehensible depth and magnitude. Unfortunately, our perception that the seas are boundless has led seafarers and others to look upon them as a receptacle for all types of garbage. For years we never saw the problem—metal and glass garbage sank, and paper and cloth decayed. But today more and more manufactured objects are made of plastics—materials renowned for their light weight, strength, and durability. The unique characteristics that have made plastics so successful, however, have made them a visible problem in the ocean.

The widespread presence of plastics in the oceans is a global problem that will require international cooperation to solve. But the roots of the problem stem from individual human carelessness in disposing of a material that is part of our everyday lives. No one can point the finger at a particular country, region, industry, or group as the major contributor to the problem. The responsibility is shared by us all.

On December 31, 1987 the United States ratified Optional Annex V of the International Convention for the Prevention of Pollution from Ships, also known as the MARPOL Protocol (pronounced MAR-POLE and short for Marine Pollution), which put an end to a centuries old practice of dumping vessel generated garbage at sea. A key factor of Annex V is its prohibition on the dumping of all plastic wastes, including plastic packaging materials and fishing gear, from all ships at sea. Not only does this mark the first effort in U.S. law to address the problem of plastic debris in the oceans, but U.S. ratification of Annex V enabled the law to come into force internationally on December 31, 1988. According to U.S. law, it is now illegal for any ship of any size to dump plastic trash in the oceans, bays, rivers and other navigable waters of the U.S.

At the same time, industry groups are convening international meetings to discuss how they can help reduce the problems caused by plastics in the

1

ocean. Environmentalists are joining with industry groups to combat the problem. And citizens across the country are getting involved in individual action from beach cleanups to lobbying.

This guide is intended to inform you—the concerned citizen, educator, researcher, or policy maker of the origins and impacts of marine debris, provide detailed information on what is being done to combat the problem, and suggest ways individuals and organizations can help. Our goal is to explain why plastic debris poses a particular threat in the ocean, and what citizens can do to reduce this threat. The future of the ocean and its resources rests with an informed public.

Chapter I introduces the reader to the problems caused by plastic debris in the marine environment. **Chapter II** examines the types of debris that are known to cause the most severe problems. **Chapter III** discusses both the ocean and land-based sources of plastic debris. **Chapter IV** examines international, federal, and state authorities governing the disposal of plastics in the oceans, particularly entanglement of marine species in discarded plastic materials. **Chapter V** details how the problems caused by plastic debris are being addressed, and **Chapter VI** suggests ways that citizens can participate in these efforts. **Chapter VII** discusses the implications and scope of the problem for the future. Finally, the appendices contain lists of agencies, people, organizations, reports, and other information for the reader who wants to know and do more. We hope you will.

I

PLASTIC DEBRIS: MORE THAN A LITTER PROBLEM

THE SUCCESS OF PLASTICS

Plastics have been in existence for more than a century, but the commercial development of today's major plastic materials occurred during World War II when shortages of rubber and other materials brought plastics into great demand. Newer plastics proved to be excellent substitutes for traditional materials such as wood, paper, metal, and glass. When large-scale plastic production commenced, reduced costs set the stage for a whole new era. By 1960, approximately 6.3 billion pounds of plastic were produced in the United States in one year. The early 1970s saw more than a three-fold increase in plastics production, totalling more than 20 billion pounds per year. In subsequent years plastic production in the United States continued to increase to the present volume of over 59.4 billion pounds in 1988: more than 10 pounds of plastic for every person on earth.

Even more notable is the growth in plastics applications. Just survey your own surroundings—bottles, bags, telephones, carpeting, toys and sports equipment, automobile parts, and boats. In fact, in most places it is difficult *not* to find something made of plastics. The convenience of plastics for consumer products and packaging is demonstrated by the fact that in 1987, the United States produced more than 34 billion plastic bottles, more than one billion pounds of plastic trash bags, and 201 million pounds of plastic for disposable diapers. Lightweight plastic has enabled fishermen to use nets that may extend over 20 miles across the ocean. Its strength has made plastic an ideal substitute for other materials used in packaging heavy cargo. For instance, plastic is rapidly replacing steel as strapping used to bind crates, and in 1987 some 450 million pounds of plastic were used in shipping sacks and pallet shrink wrap.

Major markets for plastic goods include transportation, packaging, construction, electrical and electronics, furniture and furnishings, consumer and institutional supplies, industrial parts, and machinery. In essence, plastics have been incorporated into virtually every industrial and commercial sector of America.

Growth of Plastics Production in the U.S.

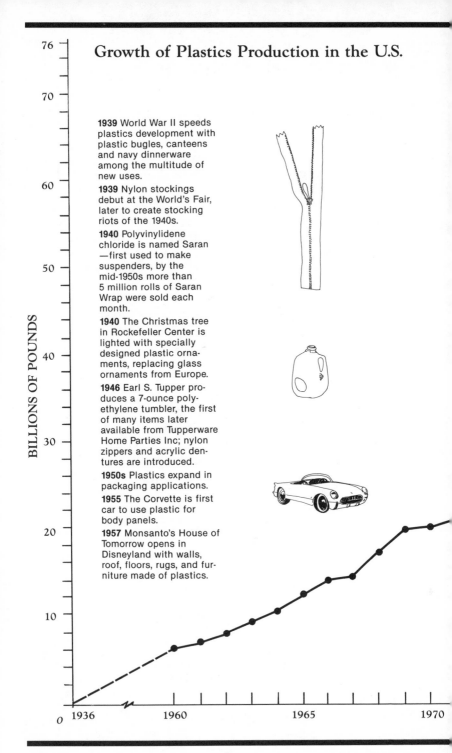

1939 World War II speeds plastics development with plastic bugles, canteens and navy dinnerware among the multitude of new uses.

1939 Nylon stockings debut at the World's Fair, later to create stocking riots of the 1940s.

1940 Polyvinylidene chloride is named Saran —first used to make suspenders, by the mid-1950s more than 5 million rolls of Saran Wrap were sold each month.

1940 The Christmas tree in Rockefeller Center is lighted with specially designed plastic ornaments, replacing glass ornaments from Europe.

1946 Earl S. Tupper produces a 7-ounce polyethylene tumbler, the first of many items later available from Tupperware Home Parties Inc; nylon zippers and acrylic dentures are introduced.

1950s Plastics expand in packaging applications.

1955 The Corvette is first car to use plastic for body panels.

1957 Monsanto's House of Tomorrow opens in Disneyland with walls, roof, floors, rugs, and furniture made of plastics.

BILLIONS OF POUNDS

76
70
60
50
40
30
20
10
0

1936   1960   1965   1970

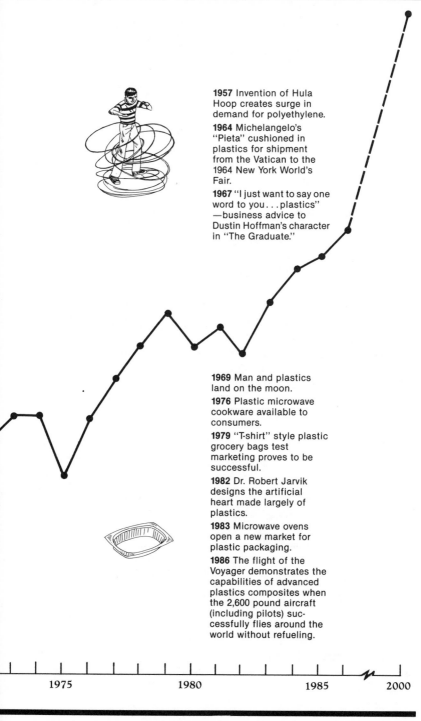

1957 Invention of Hula Hoop creates surge in demand for polyethylene.

1964 Michelangelo's "Pieta" cushioned in plastics for shipment from the Vatican to the 1964 New York World's Fair.

1967 "I just want to say one word to you...plastics" —business advice to Dustin Hoffman's character in "The Graduate."

1969 Man and plastics land on the moon.

1976 Plastic microwave cookware available to consumers.

1979 "T-shirt" style plastic grocery bags test marketing proves to be successful.

1982 Dr. Robert Jarvik designs the artificial heart made largely of plastics.

1983 Microwave ovens open a new market for plastic packaging.

1986 The flight of the Voyager demonstrates the capabilities of advanced plastics composites when the 2,600 pound aircraft (including pilots) successfully flies around the world without refueling.

1975   1980   1985   2000

BUT WHAT HAPPENS TO PLASTIC GOODS WHEN THEIR USEFUL LIFE IS OVER?

It has been common practice for crews on oceangoing vessels to throw their wastes overboard. Litter disposed of on land that is washed into marine areas via rivers and sewer systems contributes still more. Trash blown from landfills, tossed by careless beachgoers or lost from other sources adds to the problem. Most man-made debris can create aesthetic or environmental problems, but the qualities that have made plastics so successful make plastic debris a particular threat in the ocean.

Plastic is lightweight.

Plastic is used to manufacture and transport larger forms and quantities of goods without the bulk of heavier types of packaging. But plastic is also buoyant and tends to float, causing a multitude of problems in the ocean. At least 50 of the world's 280 species of seabirds are known to ingest small pieces of floating plastic, mistaking it for food such as plankton or fish eggs that also float at the surface. Five species of sea turtles are also reported to eat plastic articles, primarily plastic bags and plastic wrap, which can resemble prey, such as jellyfish, in color, shape, size, and motion. Ingested plastic may lodge in an animal's intestines and stomach, blocking its digestive tract. If large quantities of plastics are ingested, an animal may eventually die of malnutrition caused by false feelings of satiation. They may simply stop eating. For sea turtles, ingested plastics may even cause excess buoyancy, rendering a turtle unable to dive for prey or escape from predators.

Lost or discarded fishing nets and other plastic items suspended in the water column are known to entangle countless numbers of marine species. "Entanglement" refers to the interaction of a marine animal with debris that encircles its neck, flipper, tail, or other body parts. Entanglement may occur when an animal comes into either accidental or intentional contact with an item. Some of these entanglements have been attributed to the animal's inability to see plastic debris, especially fishing gear that is designed to be nearly transparent in water. Floating debris attracts fish, crabs, and other sea life which congregate under it for shelter. In fact, fishermen routinely inspect around floating debris for groups of tuna and other fish, and in some areas floating objects are intentionally placed in the water to attract these commercial species. A seal, sea turtle, or seabird attracted to prey clustered under a floating plastic object may become entangled.

In other cases, encounters with debris may be more deliberate. Young seal pups in particular are attracted to floating debris because of their curious and playful nature. Unfortunately, such curiosity may lead to entanglement.

Floating plastic, such as bags, is also a menace to navigation since plastic can disable a vessel by fouling boat propellers and water intakes. Many vessels have incurred losses of thousands of dollars to repair damages caused by plastic debris. If a vessel becomes disabled during a storm the situation could be life threatening.

Plastic is strong.

Unlike other materials that tend to dissolve, rot, tear, crack, or shatter, plastic can withstand a great degree of abuse—even wind and waves. But once a marine animal becomes entangled in a plastic strapping band, net, or other plastic item there is very little the animal can do to break free. Sea turtles entangled in fishing line have managed to swim with this burden, only to become snagged on a rock, coral, or some other bottom structure. Because of the line's strength, they are unable to break free and eventually drown. Seabirds and waterfowl such as ducks and geese that become entangled in fishing line and other debris have managed to fly, only to strangle after becoming snagged on a tree limb or power line.

Plastic is durable.

Its durability is probably its most valuable characteristic. Plastics are relatively immune to natural mechanisms of decay. Consequently, the "life" of objects made of plastic is much longer than those made of different materials. This persistent nature of plastics is the greatest threat to the marine environment—plastic debris can continue to entangle and kill marine species years after it is lost or discarded.

WHEN DID THIS PROBLEM FIRST BECOME APPARENT?

In the 1930s, researchers reported the occasional entanglement of northern fur seals on the Pribilof Islands of Alaska. These early reports documented instances of seals entangled in rubber bands cut from inner tubes, pieces of cord, string, and rawhide. Throughout the 1960s, however, entangled fur seals were noted with greater frequency and concerns grew over what looked like an increase in the number of seal entanglements in lost or discarded fishing net fragments. At the same time synthetic fiber fishing gear, or plastics, were coming into wide scale use in North Pacific commercial fisheries. In fact, by 1964, all netting material made in Japan, the major supplier of nets to American fisheries, was made of plastic. As extensive fishing operations using these nets developed in areas frequented by fur seals, incidences of entanglements increased.

In 1969 U.S. fur seal managers began to monitor the incidence of entangled seals during the commercial seal hunt. After more than a decade, the rate of fur seal entanglement had not diminished and the impact of such entanglements was receiving more attention. Studies in 1982 indicated that the ongoing decline in the North Pacific fur seal population could represent an annual mortality rate of 50,000 northern fur seals per year due to entanglement.

At about the same time, it became apparent that critically endangered Hawaiian monk seals were also becoming entangled in fishing gear and other debris, and that this could be contributing significantly to monk seal mortality. Other data indicated that lost and discarded fishing gear and other marine debris were affecting marine resources globally. Evidence that marine wildlife were swallowing marine debris added a new dimension to the problem. Others noted that not only was marine debris threatening marine mammals, sea turtles, and seabirds, but it could be having a serious economic as well as environmental impact on fishery resources due to the effects of "ghost fishing" by lost or discarded gear.

In 1984 an event took place that would become the impetus for future scientific, government, and citizen efforts to address the marine debris problem—the Workshop on the Fate and Impact of Marine Debris which was held November 27–29 in Honolulu, Hawaii. The scope of the workshop was limited to identifying the scientific and technical aspects of the marine debris problem and its impact on marine species.

But the Workshop was more than a scientific gathering. For the first time individuals met to share and discuss their own observations of the impacts of marine debris. One scientist from Hawaii had compiled worldwide documentation of sea turtle ingestion and entanglement in plastics. Another from Alaska had done the same for seabirds. A scientist from New England reported on the effects of lost gill nets. And an official with the Oregon Department of Fish and Wildlife reported on the results of Oregon's first volunteer beach cleanup.

That exchange of information inspired new studies and projects, and helped put the marine debris problem into the public spotlight.

At the 1989 Second International Conference on Marine Debris, scientists and policymakers evaluated the results from the preceding four years of marine debris work. Significant additions to the conference agenda included working group and technical sessions focused on marine debris education, economics, technology, and legislation. All participants agreed we have made tremendous progress in the last four years and we can look forward to more support from the public, government agencies, and the scientific community.

JUDIE NEILSON: A CONCERNED CITIZEN

It all began when the May/June issue of Alaska Fish and Game magazine was delivered to her office at the Oregon Department of Natural Resources by mistake. Flipping through it, she was drawn to an article entitled "The plague of plastics," that told about the increasing proliferation of plastic debris in the environment and the resulting impacts on wildlife.

Then she got an idea to organize a cleanup of plastic debris on Oregon's 350 miles of coast. She formed a steering committee and they divided the coast into 14 zones and found local residents to be "zone captains" to identify which areas were accessible and where debris, once collected, could be stacked.

Her original idea was to have 1,500 volunteers—roughly 10 for every 150 miles of accessible beach. Saturday, October 13, 1984 was selected as the cleanup day to coincide with the Year of the Ocean and Coastweek activities. The hours of 9 a.m. to 12 p.m. were chosen since they agreed with favorable tides.

Saturday morning dawned to high wind, hail, and driving rain. But despite the black sky and bleak forecast, volunteers arrived by the car and busload, dressed for the weather and raring to go.

The results: a total of 2,100 volunteers participated in the cleanup. More than half came from inland cities, driving at least 75 miles. They collected more than 26 tons, filling 2,400 20-gallon bags with plastic, including approximately 48,900 chunks of polystyrene larger than a baseball, 6,100 pieces of rope, 5,300 plastic food utensils, 4,900 bags or sheets of plastic, 4,800 plastic bottles, 2,000 plastic strapping bands, 1,500 six-pack rings, and 1,100 pieces of fishing gear.

Judie Neilson has now not only successfully generated interest in volunteer beach cleanups from Maine to Hawaii, but her idea inspired beach and village cleanups throughout the Mediterranean, in Egypt, France, Greece, Israel, Jordan, Morocco, Spain, and Turkey.

And it all started because a magazine landed on her desk by mistake.
(For more information on beach cleanups see Appendix F.)

II

LOOPS AND HOOPS AND STRAPS AND TRAPS . . .

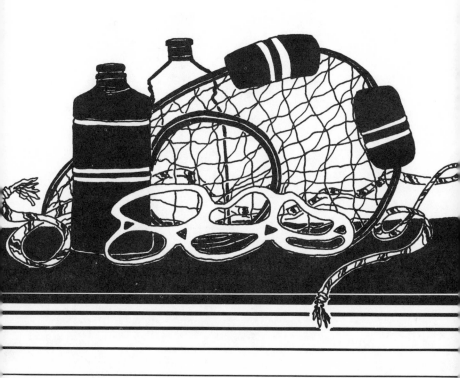

... THAT'S WHAT DEBRIS IS MADE OF

As diverse as the applications of plastics are in today's society, so are those items found in the marine environment. Plastic debris found in the marine environment generally falls into two categories: manufactured plastic articles and plastic resin pellets. Manufactured plastic articles include those items that are obvious artifacts of human activities, such as fishing gear, packaging materials, bags, and bottles. Plastic resin pellets are the raw form of plastic—typically in the shape of small spherules or beads about the size of a match head—that have been manufactured from petrochemicals and later are molded into plastic items.

Although we know enough about the plastic debris problem to draw some general conclusions about the types of plastic found in the marine environment, no one knows just how much plastic is out there. Most of the quantitative estimates of debris provide information only on isolated concentrations in relatively localized areas, such as the results from a citizen beach cleanup. But information on the amount of debris found on one beach cannot be used to estimate the total amount found in the ocean because plastics are subject to a multitude of factors including tides, winds, and currents. These ocean processes may concentrate debris in some areas, yet help keep other areas free of debris. Moreover, the types as well as the quantities of plastic debris found in an area are closely related to the identity and proximity of potential sources of debris.

For instance, in areas of Alaska adjacent to major fishing grounds, the majority of reported debris items have been pieces of lost or discarded fishing gear. In heavily populated metropolitan areas such as New York, however, wastes generated from land-based sources are most prevalent.

Therefore, we don't know the most common types of plastic debris items found in the ocean and how many are present worldwide. But we do know that plastics are now the most common man-made objects sighted at sea. In one survey, 86 percent of the trash observed floating in the North Pacific Ocean was plastic. Even in remote areas of Antarctica, researchers find plastic bottles, bags, and sheeting that have washed ashore. Some have suggested that a lack of information on total quantities of plastics in the ocean points toward the need for future studies to determine how much is out there. However, others feel that the usefulness of quantification is questionable. To determine the amount of debris in the world's oceans would require substantial effort and resources that could perhaps be better spent on solving the problem, such as developing adequate waste disposal technology for ships or reducing the escapement of plastics from land-based sources.

15

This chapter discusses the major types of marine debris known to affect the marine environment. This list is by no means exhaustive but includes debris items that are known to cause the greatest problems. These debris types have been grouped under six categories: fishing gear, cargo-associated wastes, wastes generated by offshore petroleum operations, plastic pellets, sewage-associated wastes, and domestic wastes.

FISHING GEAR

The introduction of plastics in fishing gear has been one of the most important technological advances for the modern fishing industry. Not only are plastic nets and lines lighter and easier to handle, they are also stronger, more durable, and ultimately cheaper than cotton, hemp, and other natural fibers used in the past. The demand for plastic and plastic-coated wire traps in U.S. commercial fisheries is growing because plastic is impervious to organisms that destroy wood and doesn't corrode like metal. Plastic coatings extend "trap life" for years. All-plastic traps never corrode.

Nets and Rope as Debris

Nets that are either accidentally or deliberately discarded at sea are killing marine wildlife, wasting fishery resources and endangering human safety. At one time an estimated 50,000 northern fur seals died each year due to entanglement, primarily in net fragments. Today, this number has declined to 30,000, but only because the fur seal population is smaller now. Lost nets that "ghost fish" continue to catch finfish and shellfish that are never retrieved by fishermen. One derelict gill net found off Alaska measured over nine miles in length. Entangled in the webbing were hundreds of valuable salmon and 350 dead seabirds. Nets and rope also disable vessels after becoming wrapped around propellers. Several near-fatal incidents of scuba divers entangled in lost nets have been reported.

Traps as Debris

Derelict traps made entirely or in part of plastic also compete with fishermen by "ghost fishing." Some lost lobster traps in New England, for example, are still catching six pounds of lobster per year. Although this may not seem significant, in one year an estimated 1.5 million pounds of lobster valued at $2.5 million were captured by ghost fishing traps and never retrieved. Tanner and king crab fishermen report losing six to ten percent of their pots every year. It is estimated that more than 30,000 king crab traps have been lost in the western Gulf of Alaska since 1960. Plastic net portions of lost traps that break free also entangle marine animals.

16

Plastic Debris: Signpost to Civilization?

When Steven Callahan's small sloop sank off the Canary Islands, he drifted in the Atlantic in a small five-and-a-half-foot inflatable raft for 76 days, drifting 1800 miles. For 60 days he saw practically no signs of human-kind, when finally his raft drifted into a "highway" of trash as far as he could see. Old bottles, baskets, fishnet webs, ropes, floats, and polystyrene became his "signpost to salvation." At this point he felt that food, shelter, and civilization were not far away. However, it was not until 14 days and more than 300 miles later that he spotted land. The next morning he was rescued by a fisherman.

BERRY'S WORLD / Jim Berry

© 1987 by NEA Inc

"Unfortunately, these days, seeing trash doesn't necessarily mean you're near land."

Fishing Line as Debris

For sea turtles and birds, discarded monofilament fishing line is lethal. Turtles that become entangled in line are unable to break free, and drown. One turtle found in New York had actually ingested 590 feet of heavy duty fishing line. An ornithologist in North Carolina found the body of a laughing gull entangled in fishing line. In an attempt to remove the line from the beach the man began to retrieve the remainder of the line. Twenty-five yards later he found five more birds entangled. Apparently, after the first bird became entangled it dragged the line back to its nesting colony where the line then entangled others. Ospreys, gulls, and other birds even collect pieces of line as nesting material, thus creating death traps for their young. Fishing line is also a nuisance and hazard to commercial and recreational boaters who waste time and money on damages caused when line wraps around propellers. Some boaters are even installing devices to combat this problem.

CARGO-ASSOCIATED WASTES

Plastic is being used more and more in cargo transportation. Plastic strapping, for instance, is used to bind items individually or in boxes. It has replaced rope and is rapidly replacing steel because it is lightweight, does not rust, is less dangerous when cut, and it is also about half as expensive as steel. Shipping sacks made of plastic and plastic shrink wrap for cargo pallets are also being used increasingly because of their convenience during transportation.

Plastic Strapping as Debris

Discarded plastic strapping becomes a problem when it is cast into the water, particularly when it is removed from a package without being cut, thus forming a ring that can entangle wildlife. Seals are the major victims. In fact, strapping is a common item seen on entangled seals, second only to fish netting. A synthetic "collar" of plastic strapping can cause lacerations prone to infection. As the seal grows the band will become more constricting, eventually causing strangulation.

Plastic Sheeting as Debris

Large plastic sheeting often fouls fishing gear and damages propellers, leading to vessel disablements. Sea turtles and marine mammals eat pieces of plastic sheeting they mistake for jellyfish and other prey. And fishermen, particularly those operating in the Gulf of Mexico, have voiced concerns over large pieces of plastic sheeting that become fouled in nets and seriously affect their ability to fish.

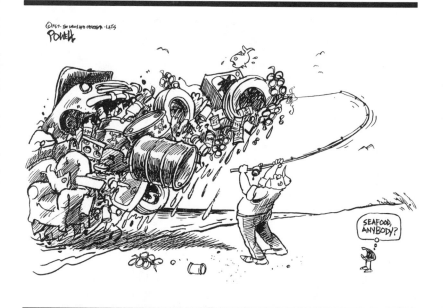

PETROLEUM INDUSTRY PLASTICS

There are certain types of debris found on beaches in the Gulf of Mexico that can be traced back to offshore petroleum industry activities. Hardhats is one. Another is 9-track "write-enable" rings. These are plastic circular rings about four inches in diameter which are used to protect tapes used during seismic recording and other computer activities. Although these types of debris pose primarily an aesthetic problem, one researcher in Texas has witnessed seabirds attempting to eat protection rings probably mistaken for food.

PLASTIC PELLETS

Plastic pellets are the raw form of plastic after it has been synthesized from petrochemicals. They are transported in bulk to manufacturing sites, where they are melted down and made into all kinds of plastic consumer goods.

Plastic pellets have been reported by researchers in many parts of the world's oceans. Although there are no estimates of the amount of pellets that escape into marine areas each year, pellets have been found in concentrations up to 3,500 per square kilometer on the surface of the Atlantic and up to

Plastic Debris: A Way Of Life

In southern latitudes deforestation and development of coastal areas has led to a decrease in the amount of natural ocean debris such as logs, coconuts, and sea beans, which historically have provided habitat for tiny ocean organisms. In its place, increasing amounts of plastic debris may be becoming an alternative way of life for some small marine animals who spend their lives attached to floating debris. One particular species of bryzoan, an animal that lives in colonies and grows like moss but within a hard shell, was once relatively rare on the Atlantic coast of Florida. Now it is the most abundant bryzoan found in the area. Researchers postulate that the recent success of this organism may be attributed to the presence of large quantities of drifting plastic in the area. It appears that this animal is able to live on plastic—where it faces no competition and no predators.

34,000 per square kilometer in the Pacific—comparable to about 50 pellets on an area the size of a football field. But although plastic pellets are not as abundant as other debris items in the ocean, in one study of Alaskan seabirds they composed about 70% of the plastic eaten. Therefore, researchers speculate that seabirds selectively choose and actively seek plastic pellets over other debris. Why?

Researchers propose that to seabirds, these plastics may resemble planktonic organisms, fish eggs, or even the eyes of fish and squid. Nearly all the plastics ingested by seabirds float at the surface where these natural prey are found. Many plastic pellets are similar in size and shape to fish eggs, small crabs, and other prey. Even the color of some pellets resemble natural prey with white, yellow, tan, and brown being most commonly ingested. Researchers in laboratories have even mistaken resin pellets for fish eggs.

SEWAGE-ASSOCIATED PLASTICS

Several items associated with municipal sewage and wastewater treatment and disposal have been identified as debris. These include plastic tampon applicators, condoms, thin pieces of plastic sheeting from sanitary napkins, and disposable diapers. During a citizen beach cleanup in New Jersey, for example, volunteers collected 650 plastic tampon applicators on a small section of coastline. Massachusetts and bordering coastal states share this same problem. In a beach cleanup on 30 miles of Cape Cod, Massachusetts, tampon applicators were found on 95 percent of the beaches surveyed. These items pose a severe aesthetic problem for coastal residents and visitors.

DOMESTIC PLASTICS

Domestic plastics are those items we use in our everyday lives including bags, bottles, lids, and a multitude of other items. Over the past decade the use of plastics in packaging has more than doubled: in 1975 nearly 5.6 billion pounds of plastics were used in packaging, in 1987 this figure increased to about 15.2 billion pounds.

Plastic Debris: A Matter Of Perspective

A Sea Grant agent recently vacationed in Belize where she came across a group of children one day on the beach. They appeared to be in search of something—perhaps shells or some other ocean treasures. Upon closer inspection she found that the children were holding plastic toys and other assorted broken bits and pieces of brightly colored plastics. When asked what they planned to do with this debris the children responded that they came to the beach regularly to see what gifts the ocean had brought to their shoreline. They would later take these gifts back to their families.

Plastic Bags and Sheeting as Debris

During a 150-mile survey of North Carolina beaches, more than 8,000 plastic bags were found in three hours. Bags and sheeting are ingested by marine wildlife that mistake these items for food. One turtle was found with 15 bags in its stomach, a whale was found with 50. Plastic bags also endanger human safety at sea since they can cause boat engine failure by clogging cooling water intakes.

Six-pack Rings as Debris

Plastic rings used to carry beverage cans are a threat to all kinds of marine animals. Researchers have found fish, birds, and even a California sea lion entangled in plastic six-pack rings. Along 300 miles of Texas coastline, more than 15,600 six-pack rings were found in three hours.

While the plastic objects identified above are causing the most serious problems at sea, all plastics should be disposed of properly. Even bottles and other everyday items have the potential to break into fragments which can be ingested by marine wildlife, foul fishing gear, and simply add to the accumulation of plastics in the world's oceans. One juvenile hawksbill sea

21

turtle found on a Hawaiian beach had ingested an eight-inch square plastic bag. In addition the turtle had ingested a golf tee, shreds of bag and sheeting, pieces of monofilament fishing line, a plastic flower, part of a bottle cap, a comb, chips of polystyrene, and dozens of small round pieces of plastic. In total, the sea turtle that weighed less than 12 pounds had ingested 1.7 pounds of plastic. This is comparable to 20 pounds of plastic in the stomach of a 120-pound person.

"WHAT GOES UP. . . ."

In 1985, a young sperm whale was found dying on the shores of New Jersey as the result of a mylar balloon lodged in its stomach and three feet of purple ribbon wound through its intestines. In 1987, a necropsy of a half ton leatherback turtle revealed that its digestive tract was blocked by a latex balloon and three feet of ribbon. Much attention has been focused upon plastic trash in the oceans and its danger to wildlife, yet most people fail to recognize that balloon releases contribute to the problem. There is really no difference between releasing 250,000 balloons and tossing 250,000 plastic bags over the side of a ship.

"What goes up, must come down" is an all too familiar cliche, but it's true. The beautifully colored balloons released into the air, often in large quantities, only *seem* to disappear. These balloons deflate and fall back to earth. Scientists have found that sea mammals, sea turtles, fish, and birds ingest the floating balloons mistaking them for authentic food such as jellyfish. Dr. Peter Lutz, a scientist at the University of Miami, is currently studying the ingestion of plastic and latex by sea turtles. Dr. Lutz observed that green and loggerhead sea turtles actively consumed plastic and latex materials when they were offered and found that there are increasing records of plastic and latex materials being found in the intestines of dead animals.

For many people, it is simply a case of ignorance. They are unaware that balloons pose a threat to our marine environment. Education will be a key element to eliminate the problems resulting from balloon releases. Many groups now work to inform other sponsors of balloon launches about their threat to marine wildlife and to encourage alternatives.

There is no need to take the fun and excitement out of balloon launches. Some groups have held balloon launches in covered areas such as gymnasiums or churches. A hot air balloon can generate more attention than numerous small ones, and some organizations have even attached lines to the balloons so that they could be recovered. Now that the evidence is available, people must see balloon releases as a dangerous and unacceptable form of plastic pollution.

III

WHERE DOES IT ALL COME FROM?

C ertain types of plastic debris can be easily traced to a particular source. Others, such as plastic bags and bottles, may be generated from several different and sometimes untraceable sources. But there are many sources both on land and at sea that are known to contribute to the marine debris problem.

OCEAN SOURCES

More than a decade ago, the National Academy of Sciences estimated that ocean sources dumped 14 billion pounds of garbage into the sea every year—more than 1.5 million pounds per hour. This figure includes all solid cargo and crew waste material (paper, glass, metal, rubber, and plastics) that were assumed to be disposed of by the world's commercial fishing and merchant shipping fleets, passenger cruise liners, military vessels, oil drilling rigs and platforms,. recreational boaters, vessel accidents, and major storms where substantial amounts of debris could be washed to sea.

According to the Academy, not only is the majority of ocean litter primarily concentrated in the Northern Hemisphere, but the United States could be the source of approximately one-third of all the trash in the world's oceans. Hence, a reduction in the amount of litter generated by the United States would contribute significantly to a worldwide reduction.

More than 85 percent of this trash, or 12 billion pounds, was estimated to come from the world's merchant shipping fleet in the form of cargo-associated wastes including dunnage, pallets, wires, and plastic covers. Other sources and the estimated amount of litter generated by these sources each year are given in the table on the next page.

Of the total amount it was estimated that 0.7% of the crew litter, or more than 8 million pounds per year, was plastic for merchant, commercial fishing, and military vessels. It was assumed that on passenger vessels 1.8% of all litter, or more than 1 million pounds, was plastic since these vessels operate somewhat like floating hotels and cater to the public. The percentage of plastics for recreational vessels and oil rigs and platforms was not given, and these figures do not take into account the amount of plastic fishing gear and cargo wastes dumped at sea.

But the use of plastics has increased since the time of this study. At present, it is estimated that 20 percent of the value of all food packaging material in the United States is plastic and that this figure may increase to 40 percent by the year 2000. For instance, metal cans and glass jars and bottles are major targets for expanding the use of plastics. Although it may seem that plastic containers already dominate the market, they composed only twenty percent of all rigid containers in 1987 (59 percent were metal and 21 were glass).

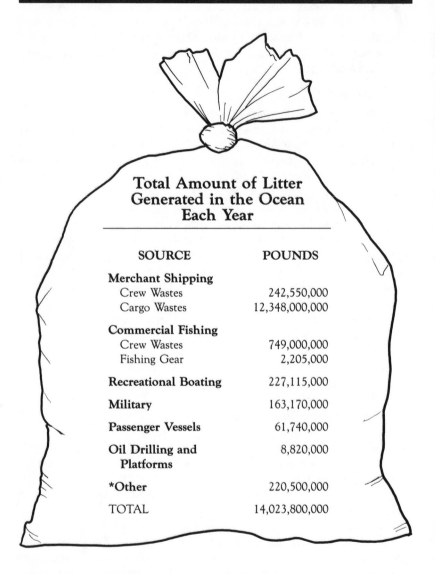

Total Amount of Litter Generated in the Ocean Each Year

SOURCE	POUNDS
Merchant Shipping	
Crew Wastes	242,550,000
Cargo Wastes	12,348,000,000
Commercial Fishing	
Crew Wastes	749,000,000
Fishing Gear	2,205,000
Recreational Boating	227,115,000
Military	163,170,000
Passenger Vessels	61,740,000
Oil Drilling and Platforms	8,820,000
***Other**	220,500,000
TOTAL	14,023,800,000

* This includes debris from shipping accidents and from major storms in coastal areas.

Adopted from National Academy of Science, 1975

Although more recent quantitative estimates are not available, the disposal of wastes from ocean sources has continued essentially because it is inexpensive and convenient. Alternative means of handling shipboard wastes such as incinerators, grinders, and compactors are costly. Small vessels do not have the space for large waste-processing equipment. For larger vessels, such equipment is often either not durable enough to handle shipboard shock and vibration, or is considered to be a potential safety hazard, as with incinerators. Furthermore, vessels that store garbage on board require adequate facilities on shore for disposal. But, prior to implementation of Annex V, many ports both in the United States and abroad did not have such facilities. In the United States, the matter of waste disposal for ships is complicated by the fact that ships entering U.S. waters from destinations outside of the United States and Canada must incinerate, sterilize, or grind and deposit into an approved sewage system any garbage that has at one time been in contact with food (including plastic packaging) and disposed of at a facility that has been approved by the U.S. Department of Agriculture. These regulations are designed to prevent the introduction of foreign agricultural pests and diseases like swine fever and hoof-and-mouth disease into the United States. But there has been some question as to the availability, convenience, and economics of these facilities. Consequently, in many situations garbage, including plastics, was routinely dumped overboard before the vessel entered port. In fact, of the 73,600 vessels that entered U.S. ports in 1986, fewer than 1,900 or 3 percent had garbage to offload.

More Recent Information on Ocean Sources of Debris Prior to Annex V:

Merchant Shipping Industry: The world's fleet of merchant vessels dumped at least 450,000 plastic containers, as well as 4,800,000 metal and 300,000 glass containers, into the sea every day.

Commercial Fishing Industry: The world's commercial fishing fleets discarded more than 50 million pounds of plastic packaging and 100,000 tons of plastic fishing gear including nets, rope, traps, and buoys every year.

United States Navy: The United States Navy has typically thrown all wastes overboard. Each crew member generates more than three pounds of solid waste per day, up to .2 lbs. of which is plastic. Some ships having 5,000 men onboard can generate 1,000 pounds of plastic trash per day.

Passenger Ships: Before Annex V, passenger ships serving U.S. ports alone have jettisoned nearly 62 million pounds of litter each year. Although passengers have little say in the matter of garbage disposal on a cruise liner it is regrettable to note that because approximately 70 percent of all passengers on cruise liners are U.S. citizens, a large portion of the trash generated by passenger ships comes from Americans.

Recreational Vessels: More than nine million recreational vessels are registered in the United States. They have dumped more than 100 million pounds of garbage into U.S. coastal waters every year. The greatest concentrations of recreational vessels in the United States are found in the waters of New York, New Jersey, the Chesapeake Bay, Florida, and the Great Lakes.

Recreational fishermen are also a major source of debris in the form of monofilament fishing line. The amount of fishing line lost or discarded by 17 million U.S. fishermen during an estimated 72 million fishing trips in 1986 is not known, but if the average angler snares or cuts loose only one yard of line per trip, the potential amount of deadly monofilament line is enough to stretch around the world. During the 1988 Florida coastal and inland waterways cleanup, 11,000 volunteers collected 304 miles of monofilament fishing line in just three hours.

Petroleum Industry: Although the disposal of wastes from oil rigs and drilling platforms is strictly regulated, trash associated with petroleum activities is still prevalent in coastal areas, particularly in the Gulf of Mexico. Items include buckets, hardhats, and other equipment used on the job. During a Texas beach cleanup in 1987, 225 hardhats and 2,337 "write-enable" rings were found on 157 miles of coastline.

LAND-BASED SOURCES

In some cases plastic debris generated by land-based sources is easily traced. In other cases the sources of certain debris items are not as clearly identified. Untold quantities of plastic enter the ocean via rivers, drainage systems, estuaries, and other avenues. The following provides an overview of some land-based sources of plastic including plastic manufacturing and processing plants, sewage systems, landfills, dock and marina structures, and littering by the general population.

Plastic Manufacturing and Processing Activities

While the effects of plastic resin pellets on the marine environment are receiving increasing attention, the means by which pellets enter marine systems is as yet unclear. Studies in the early 1970s reported that pellets were found in sediment samples taken within rivers below U.S. plastics factories, suggesting that plastics are directly discharged into river systems by these industries.

Current investigations thus far suggest that the prime source of pellets are transportation related incidents. Pellets presently found at sea could be a residual of a problem that once existed for the plastics industry but has since been corrected, at least in the United States, through better equipment and procedures. However, plastic pellets are still sighted in coastal areas of the United States. Until this problem is carefully assessed, it is impossible to determine whether pellet escapement is still a problem at the manufacturing level, or whether pellets enter marine areas during transportation and handling.

Industry Initiative

Dow Chemical's Louisiana Division has produced a four-minute video on its pellet reclamation procedures. With the installation of collection traps, precautions in handling, and thorough freight car cleansing procedures, Dow's Louisiana plant reclaims 500 pounds, or approximately five million pellets, each day that would otherwise escape into marine areas. For more information contact Dow Chemical Company (see Appendix I).

Sewage Operations

In the past, sewage and wastewater treatment and disposal systems also dumped plastic wastes into the oceans. In some areas sewer systems discharged plastic tampon applicators, diapers, and other plastic items directly into marine areas. Legislation amending the Marine Protection Research and Sanctuaries Act makes it unlawful for New York and New Jersey to dump sewage sludge and industrial waste into marine areas after December 31, 1991.

During a heavy rainfall, sewer systems combined with stormwater runoff can also generate marine debris. These systems, referred to as combined sewer overflows (CSO), can overflow into marine areas, thereby bypassing sewage treatment.

Solid Waste Disposal Practices

Land-based solid waste disposal sites are another source of marine debris. Contrary to a popular belief, trash is no longer loaded onto barges and legally dumped at sea in the United States. However, in some areas, garbage is emptied at collection sites onto barges and then transported to landfills located along coastal waterways. An example is Fresh Kills landfill on Staten Island, New York, which receives 26,000 tons of trash a day of which 14,000 tons is transported by barges each day. But lightweight litter such as plastic is frequently blown off the barges and into the water. Escapement into surrounding waters also occurs as a result of sloppy barge unloading. However, operations have shown continuing improvement with the use of skimmer vessels to collect garbage at loading and unloading docks, barge covers, garbage loading height restrictions, and booms on and around loading areas.

Degradation of Docks and Marinas

Large chunks of polystyrene foam are used for flotation in dock and marina structures. But they are also found as debris in marine areas. During

JAY CRITCHLEY: ARTIST

Draped in a gown rustling with 3,000 pink and white plastic tampon applicators, Jay Critchley attended the 1986 centennial celebration of the Statue of Liberty. His Miss Liberty costume, complete with a seven-pointed crown and a torch, was created totally from plastic tampon tubes that washed ashore in New Jersey and on Cape Cod, Massachusetts.

The overwhelming prevalence of plastic tampon applicators on Cape Cod and surrounding beaches has caused much consternation for local residents who complain that hundreds of applicators routinely wash up on town beaches. People have even jokingly named them "beach whistles." Jay Critchley, a local artist, has formed the Tampon Applicator Creative Klub International (TACKI) to draw attention to the problem. He collects tampon applicators found on beaches and creates sculptures.

Legislative bills have been introduced in both Massachusetts and New Jersey that would ban the sale and distribution of plastic tampon applicators, stating that whoever sells or distributes disposable tampon applicators composed of plastic or other non-degradable material shall be punished by fines between $1,000 to $5,000 for each offense.

According to Mr. Critchley, the words of the Emma Lazarus poem engraved on the statue's base, "the wretched refuse from your teeming shore," no longer refer to the metaphorical downtrodden masses but, literally, to trash.

For more information on TACKI see Appendix I.

a beach cleanup of Oregon's coast nearly 50,000 chunks of polystyrene foam larger than a baseball were collected. High concentrations of polystyrene foam at the mouths of rivers with moorages upstream suggest that docks and marinas are the most likely sources of such debris.

Littering by the General Population

People who visit the beach for recreation also contribute to the problem, leaving behind items that either remain as coastal debris, or are easily transported offshore, adding to the litter in the sea. In Los Angeles County, California, beachgoers leave behind approximately 75 tons of trash each week.

Plastic Debris: A New Science

Professor Anthony Amos, an oceanographer with the University of Texas Marine Science Institute at Port Aransas, has been involved in a long-term study of Mustang Island beach in Texas. When he first saw Mustang Island he was appalled by the amount of trash on it. Ten years later, Professor Amos is now an authority on the types of trash found on Texas beaches. But the debris has given him much information on the seasonal variation of circulation patterns in the Gulf. He knows that currents travelling parallel to the Texas coastline are coming from the south when green Mexican bleach bottles become prevalent on the beach. In late spring and fall plastic containers bearing the names of supermarkets in Louisiana, Mississippi, and Alabama indicate that the currents have changed and are from the north. By studying these debris indicators, Amos hopes to achieve a greater understanding of the current system off the Texas coast.

Researchers have noted that beaches cleaned on a regular basis are often more popular than those which are not, even though they may be highly polluted areas. In essence, the perception of pollution is often measured more by the debris seen in the water or on shore as opposed to unseen human health hazards such as bacteria, viruses, and chemicals. Others have noted that once an area appears visibly polluted by debris, people are more likely to leave their own trash behind.

IV

THERE OUGHTA BE A LAW!

L itter-strewn beaches, clogged sewer systems, fouled boat propellers—
these are the immediately comprehensible problems. Less visible are
the entanglement and drowning of marine animals in plastic debris,
the starvation and death of animals which eat plastic junk mistaking it for
food, and the waste of valuable fishery resources that are captured by lost and
abandoned fishing gear, only to rot in the sea.

The first reaction of most people, when confronted with this list of the
harmful effects of plastic debris in our oceans, is: "There oughta be a law!"

Even though concern about marine pollution in general has been
expressed since the 1970s, plastic debris in the marine environment is a
relatively new concern. Congress has recently begun to address the problem
of plastic debris, and at the close of 1987, took two important actions to target
the problem. The U.S. Senate approved the ratification of an international
agreement to ban the dumping of plastics at sea, and the Congress enacted
domestic legislation prohibiting ships from dumping plastics in the U.S. waters.
Prior to these efforts, existing legal authorities to address ocean pollution were
not plastic-specific, and could not be used to get at the particular problem of
entanglement.

OCEAN DUMPING LAWS

The first type of law concerned citizens and industries have looked to as
a way to address the plastic debris problem are laws that deal with ocean
dumping. But these laws, many of which are international in scope, often are
limited to oil spills, or other substances considered "toxic" or "hazardous."
Plastic does not fall into either of these categories. In some cases, plastic
fishing gear is specifically exempted from dumping controls if it is lost during
the course of fishing operations—a major source of plastic debris.

Annex V of the MARPOL Protocol is the newest legal tool to reduce
plastic debris. Annex V went into effect on December 31, 1988 and to date,
39 countries have agreed to abide by its requirements (see Appendix L).
Annex V prohibits the disposal of all plastics into the ocean and requires that
all vessels carry their plastic trash into port for placement in proper disposal
facilities. The kinds of trash regulated by Annex V are listed in the following
table.

Besides giving its nod of approval to Annex V, the U.S. Senate passed
a measure that implements Annex V provisions in our waters. The Marine
Plastic Pollution Research and Control Act, Public Law 100-220, combines
ideas of several different pieces of plastic pollution legislation that were introduced in 1986 and 1987. After passing the House and Senate (December 18

MARPOL Annex V
Summary of At-Sea Garbage Disposal Limitations

Garbage Type	All Vessels		Offshore Platforms & Assoc. Vessels***
	Outside Special Areas	**In Special Areas	
Plastics—includes synthetic ropes and fishing nets and plastic garbage bags	Disposal prohibited	Disposal prohibited	Disposal prohibited
Floating dunnage, lining, and packing materials	>25 miles off shore	Disposal prohibited	Disposal prohibited
Paper rags, glass, metal bottles, crockery and similar refuse	>12 miles	Disposal prohibited	Disposal prohibited
Paper, rags, glass, etc. communited or ground*	>3 miles	Disposal prohibited	Disposal prohibited
Food waste communited or ground*	>3 miles	>12 miles	>12 miles
Food waste not communited or ground	>12 miles	>12 miles	Disposal prohibited
Mixed refuse types	****	****	****

* Communited or ground garbage must be able to pass through a screen with mesh size no larger than 1 inch.

** Special areas are the Mediterranean, Baltic, Red and Black seas areas, and the Gulf's areas.

*** Offshore platforms and associated vessels includes all fixed or floating platforms engaged in exploration or exploitation of seabed mineral resources, and all vessels alongside or within 500m of such platforms.

**** When garbage is mixed with other harmful substances having different disposal or discharge requirements, the more stringent disposal requirements shall apply.

MARPOL Annex V
Sticker Produced for Boaters

It is illegal for any vessel to dump plastic trash anywhere in the ocean or navigable waters of the United States. Annex V of the MARPOL TREATY is a new International Law for a cleaner, safer marine environment. Each violation of these requirements may result in civil penalty up to $25,000, a fine up to $50,000, and imprisonment up to 5 years.

Outside 25 miles
ILLEGAL TO DUMP
Plastic

12 to 25 miles
ILLEGAL TO DUMP
Plastic
Dunnage (lining & packing materials that float)

3 to 12 miles
ILLEGAL TO DUMP
Plastic
Dunnage (lining & packing materials that float) also if not ground to less than one inch:
Paper   Crockery
Rags   Metal
Glass   Food

U.S. Lakes, Rivers, Bays, Sounds and 3 miles from shore
ILLEGAL TO DUMP
Plastic & Garbage
Paper   Metal
Rags   Crockery
Glass   Dunnage
Food

State and local regulations may further restrict the disposal of garbage.

WORKING TOGETHER, WE CAN ALL MAKE A DIFFERENCE!
CENTER FOR MARINE CONSERVATION   1725 DeSales Street, NW   Washington, DC 20036   (202) 429-5609

and 19 respectively), the President signed it into law on December 29, 1987 and it went into effect on December 31, 1988 after a twelve month period of preparation.

CONGRESSIONAL HISTORY OF THE MARINE PLASTIC POLLUTION RESEARCH AND CONTROL ACT

The following pieces of legislation all contributed to the final version of the Marine Plastic Pollution Research and Control Act of 1987 (MPPRCA).

On June 25, 1986 Senator John H. Chafee (R-RI) introduced the first bill to address the plastic debris problem, S. 2596, the *Plastic Waste Reduction Act of 1986*. The bill called for the Environmental Protection Agency to head an interagency review of the adverse effects of plastic debris in the terrestrial, marine, and freshwater environments; develop recommendations on ways to reduce or eliminate the problems; and require that within 18 months of enactment, plastic rings used for carrying beverages be made degradable nationwide. Subsequently, the following bills were introduced:

- On June 26, 1986, Senator Ted Stevens (R-AK) introduced S. 2611, the *Driftnet Impact Monitoring, Assessment, and Control Act of 1986* which would establish a bounty system for persons who retrieved lost or discarded netting for disposal in port; require a study to develop recommendations for establishing a driftnet marking, registry, and identification system to determine the origin (by vessel, if possible) of lost, discarded, or abandoned driftnets; and require an evaluation of the feasibility of using degradable materials in a portion or all of a driftnet to accelerate decomposition if a net was lost or abandoned, thereby minimizing the hazards to marine resources. A companion bill, H.R. 5108, was introduced by Representative Charles Bennett (D-FL).

- On August 11, 1986 Representative William J. Hughes (D-NJ) introduced H.R. 5380, the *Plastic Waste Study Act of 1986*. The bill directed the Environmental Protection Agency and the National Oceanic and Atmospheric Administration to conduct a joint 18-month study of the adverse effects of plastic debris, including but not limited to raw plastic pellets, six-pack rings, strapping bands, and fishing gear on terrestrial and aquatic environments, and to make recommendations for eliminating or lessening such adverse effects.

- On August 13, 1986 Representative Leon E. Panetta (D-CA) introduced H.R. 5422, which was identical to Senator Chafee's bill.

On August 12, 1986 the House of Representatives Committee on Merchant Marine and Fisheries Subcommittee on Coast Guard and Navigation conducted oversight hearings to examine the extent of plastic pollution in the marine environment and to discuss possible solutions to this problem. The hearing was chaired by Congressman Gerry E. Studds (D-MA). Testimony was given by the U.S. Coast Guard, National Marine Fisheries Service, United States Marine Mammal Commission, the Society of the Plastics Industry, environmental organizations, and representatives of the fishing industry.

In early 1987, the White House transmitted Annex V of MARPOL to the Senate for advice and consent to ratify, and members of the 100th Congress introduced more bills to address the marine debris problem, including measures that would implement Annex V in the United States. In total, eight bills were proposed in the Senate and House. These included:

- Sen. Chafee's (R-RI) *Plastic Waste Reduction and Disposal Act of 1987* (S. 559) which was different from the 1986 version in that it would require a study to examine ways to *reduce* plastic waste on land and in the oceans, including an analysis of the use of degradable plastics in fishing gear, six-pack holders, strapping bands, and other finished products that threaten fish and wildlife. It would also ban the use of nondegradable six-pack yokes.

- Sen. Chafee's *Implementation of the Provisions of Annex V to the International Convention for the Prevention of Pollution from Ships as Modified by the Protocol of 1978* (S. 560). In order for Annex V to take effect each country must enact a law that contains the provisions of the Annex. This bill would do just that.

- Sen. Frank Lautenberg's (D-NJ) *Plastic Pollution Control Act of 1987* (S. 633). The bill would prohibit the disposal of plastic products in U.S. waters, require a study to determine ways to eliminate plastic pollution, require a public awareness program about plastic pollution, and would also establish legislation to implement Annex V of the MARPOL Protocol like Senator Chafee's bill.

- Sen. Stevens' (R-AK) *Driftnet Impact Monitoring, Assessment, and Control Act of 1987* (S. 62). The bill did not differ from the 1986 version.

- Rep. Charles Bennett's (D-FL) *Driftnet Impact Monitoring, Assessment, and Control Act of 1987* (H.R. 537)—identical companion bill to S. 62.

39

PHOTODEGRADABLE SIX-PACK RINGS

On January 1, 1977 Vermont became the first state to enact legislation requiring that all plastic six-pack rings sold in the state be made of degradable plastic. These rings are photodegradable, meaning that if a plastic six-pack ring was discarded improperly ultraviolet rays from the sun would cause it to break apart. This is accomplished by changing the molecular structure of the plastic.

The photodegradable carrier was developed in the late 1970s to address two problems: litter and entanglement. The break-up of photodegradable six-pack rings starts with the sun. As the plastic breaks down, wind and rain cause the carrier to become brittle and continue to break down into smaller and smaller pieces. The amount of time for this process varies from one area of the country to another and from season to season, but in general it takes less than three months.

To date, 18 states have enacted legislation that require degradable six-pack rings. The dates of enactment are as follows:

Alaska	10/1/81	Minnesota	1/1/89
California	11/1/82	New Jersey	4/21/86
Connecticut	10/1/84	New York	9/12/83
Delaware	1/15/83	Oregon	9/1/78
Florida	7/1/89	Pennsylvania	10/1/88
Iowa	7/1/89	Rhode Island	7/1/87
Maine	1/1/78	South Dakota	7/1/90
Massachusetts	1/17/83	Vermont	1/1/77
Michigan	6/1/89	Wisconsin	1/1/90

Typically, the requirement is stated in language such as:

"No beverage shall be sold or offered for sale in this State in metal containers connected to each other by a separate holding device constructed of plastic rings or other device of material which is not degradable. Degradable means decomposition by photodegradation, chemical degradation or bio-degradation within a reasonable period of time upon exposure to the elements."

In February 1986, a bill was introduced in Congress by Senator John Chafee of Rhode Island to address the plastic debris problem. It included a requirement that all six-pack rings sold in the United States be degradable. The bill was modified in the Senate and reintroduced the following term. The law requiring degradable six-pack rings was enacted in October 1988. However, the law does not go into effect immediately. The Environmental Protection Agency has 24 months from enactment of the law to establish regulations including the acceptable period necessary for the plastic to degrade.

Anheuser-Busch Inc., the world's largest brewer, voluntarily changed all of its six-pack rings to degradable plastic in 1987. This conversion was the first by any brewer. Close on this track is the Outboard Marine Corporation which has begun packaging motor oil in photodegradable loop carriers.

To determine whether a six-pack ring is degradable look for a diamond embossed on the ring in the area adjacent to the finger hole as shown below. This tells canners, retailers, and consumers that the carrier is degradable and meets the specifications of state law.

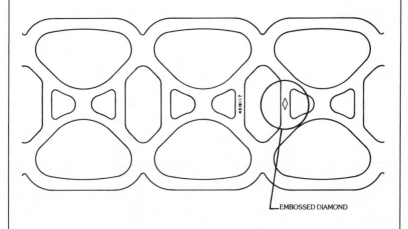

EMBOSSED DIAMOND

Illustration courtesy of Illinois Tool Works

- Reps. Hughes' (D-NJ) and Claudine Schneider's (R-RI) *Plastic Waste Study Act of 1987* (H.R. 474). The bill required a study and quantification of the overall effects of plastics on marine animals, and made recommendations to eliminate or lessen such effects through several means, such as use of degradable plastics and recycling.

- Rep. Studds' (D-MA) *Plastic Pollution Research and Control Act* (H.R. 940). The bill would also establish enabling legislation to implement Annex V of the MARPOL Protocol, require development of a plan to deal with portside reception of garbage, require negotiations with foreign governments whose fishermen take U.S. marine resources with gill nets to monitor impacts on marine mammals and reduce unintended injury and death, identify the magnitude of driftnet impacts on marine animals, evaluate driftnet marking and registration, examine driftnet degradability, establish a seabird protection zone around the Aleutian Islands, and quantify the overall effects of discarding plastics into the marine environment.

- Rep. Panetta's (D-CA) *Plastic Waste Reduction Act of 1987* (H.R. 2527) which would direct the Environmental Protection Agency to conduct a study of the possible solutions to certain problems associated with plastic debris including degradability and to control pollution by plastics on land and in the water.

On April 2, 1987, 30 U.S. Senators sent a letter to the President requesting "assistance in developing a coordinated strategy to resolve the increasingly serious and complex problems resulting from the presence of plastic debris in the marine environment." Major points included the need to "focus and coordinate the various efforts of the federal government to develop solutions to the marine plastic debris problem." The letter called for the establishment of an interagency task force that would set forth an action plan of measures to reduce the marine debris problem as well as research and development efforts and additional legislation as warranted. The National Oceanic and Atmospheric Administration was designated to chair the interagency group and outlined a six to nine-month study to assess the problem and need for research, identify potential measures to reduce marine debris, and consider alternative actions. The group included the Department of Agriculture, Department of Interior, Department of State, Coast Guard, Navy, the Environmental Protection Agency, the Council on Environmental Quality, the Office of Management and Budget and the Domestic Policy Council.

By the fall of 1987, hearings had been held on the Studds, Chafee,

42

Lautenberg, and Stevens bills, the Administration submitted its own version of an implementation bill, and the Senate Foreign Relations Committee recommended that the Senate ratify MARPOL Annex V. U.S. ratification would provide the necessary shipping tonnage (50 percent of the world's fleet) to bring this treaty into force internationally within the next 12 months.

On October 13, 1987, the House voted 368 to 14 in favor of H.R. 940, introduced by Representative Studds. This bill set out measures to implement Annex V prohibitions in U.S. waters. Meanwhile in the Senate, substantial negotiations in the Senate Environment and Public Works Committee worked out differences between the Lautenberg and Chafee bills. Further compromise with the Commerce Committee led to a Senate version which also described

GUMDROP

"WELL, IT **DOES** SEEM MORE LIKE THE BEACH NOW, BUT THIS WASN'T EXACTLY WHAT I HAD IN MIND..."

Annex V implementation. The final legislation that passed incorporated H.R. 940 into H.R. 3674, a catch-all measure passed by the House on December 18. The Senate approved H.R. 3674 on December 19 and it was signed by the President on December 29, 1987 (P.L. 100-220). H.R. 3674 included authority to implement the provisions of Annex V; to assess and mitigate the negative effects of plastics in the environment; and to improve efforts to monitor, assess, and reduce the adverse effects of driftnets.

Specifically the MPPRCA prohibits the disposal of plastics at sea by U.S. vessels effective December 31, 1988, the date on which Annex V enters into force in the United States. Additionally, it prohibits the disposal of plastics by any vessel within U.S. waters, including bays, sounds and inland waterways, out to 200 miles. The law also regulates disposal of non-plastic items depending on the vessel's distance from shore.

Unfortunately, most laws governing ocean-based activity have major enforcement problems. Since it is impossible to patrol ocean waters, which cover more than two-thirds of the earth's surface, it is necessary to create incentives to comply with the law, and to provide vessel waste handling and reception facilities that are both easily accessible and economically practical. Such strengthening measures were incorporated into the MPPRCA.

Requirements of the 1987 Marine Plastics Pollution Research and Control Act

Placards and Logbooks

The law authorizes the Coast Guard to write rules and regulations regarding the display of placards to notify crew and passengers of the requirements of Annex V. It also requires each vessel to keep a log book on garbage disposal. (A similar log is required for oil under regulation 20 of Annex I of the MARPOL Protocol.) An entry in the log is required each time vessel garbage is offloaded at port, or handled onboard by incineration or some other means.

Waste Management Plan

All U.S. vessels must develop and use a shipboard waste management plan specifying how it plans to comply with the provisions of Annex V. In addition, while Annex V applies only to ships of countries which are signatory to the MARPOL Protocol, the law gives the Coast Guard additional authority to prosecute any vessel operator who dumps plastics within 200 miles of the U.S. coast.

44

Compliance of Public Vessels

Under current international and domestic law, public vessels are exempt from MARPOL restrictions, but the MPPRCA directs all federal agencies, including the Navy and Coast Guard, to bring their vessels into full compliance with Annex V regulations by 1994.

Reception Facilities at Ports and Terminals

The Department of Transportation must make sure that ports and terminals have adequate facilities for collecting shipboard wastes. The law requires all marinas, yacht clubs, fuel docks, etc.

Citizen Monitoring

One element of the law that encourages citizen aid in enforcement is a reward provision. This allows the courts to award a portion of criminal penalties or civil fines assessed against the violator to the person who gave the information that led to a conviction or assessment of a penalty. Under the MPPRCA, it is even possible for private persons to bring actions against violators, or against the government if enforcement agencies fail to carry out their responsibilities under the law.

Study of Methods to Reduce Plastic Pollution

The Environmental Protection Agency (EPA), in consultation with the National Oceanic and Atmospheric Administration (NOAA), is directed to study the effects of plastics on the marine environment and plastics in the solid waste stream. Further, the EPA must study plastic waste and other forms of pollution in the New York Bight area, including an assessment of land-based sources of plastics and municipal sewage sludge dumping practices. A report to Congress will be available in June 1989 and a restoration plan for the New York Bight is due by 1990.

Research on Effects of Plastics on the Marine Environment

The law also required a report by NOAA to Congress which will be available in August 1989 through NMFS office of Protected Resources. The report must include identification of harmful plastics in the marine environment, their effects on living marine resources, an analysis of degradable plastics, and recommended legislation to prohibit, tax, or regulate sources of plastic materials which enter the marine environment.

Public Education

The law also requires NOAA and EPA in consultation with the Coast

Guard to conduct a three-year public outreach program to educate boaters, fishermen, and other marine user groups about the harmful effects of plastic pollution, and the need to reduce such pollution and to recycle plastic materials.

Special Areas

By international agreement the MARPOL regulations designate particular ocean regions as "special areas." Specific oceanographic and environmental characteristics cause these areas to be more susceptible to pollution by garbage. In "special areas" the law prohibits dumping of all materials regardless of the distance from shore. The difference between Annex V prohibitions in waters generally, as compared to waters in designated special areas, is illustrated in the table on page 36. The law allows ships to dump food wastes when they are further than twelve nautical miles from shore. Currently the following regions are designated as "special areas": The Mediterranean, the Baltic Sea, the Black Sea, and the Middle Eastern Gulf areas. Of these, "special area" restrictions designated as "special areas": The Mediterranean, the Baltic Sea, the Black Sea, and the Middle Eastern Gulf areas. Of these, "special area" restrictions will only be enforced in the Baltic Sea. "Special area" dumping restrictions will not be enforced in any other areas until the regions can demonstrate proper reception facilities and subsequently satisfy a one-year waiting period.

The Driftnet Impact Monitoring, Assessment, and Control Act of 1987

Another issue addressed in Public Law 100-220 under Title IV is the effect of lost or abandoned driftnets. The purpose of this act is to assess and minimize the adverse effects of driftnets in the marine environment. NOAA was directed to arrange for cooperative international monitoring and research programs with foreign countries that conduct high seas driftnet fishing operations in the North Pacific Ocean to assess the impacts of driftnets on marine resources. The law not only addresses the incidental take of marine animals during driftnet fishing operations, but also the problems caused by lost or discarded driftnets. (Since accidental loss of plastic fishing gear is not covered by Annex V, it is difficult to distinguish accidental losses from deliberate disposals of unwanted gear.)

Title IV directed NOAA to evaluate the feasibility of establishing a driftnet marking, registry, and identification system to provide a reliable method for identifying the origin of lost or abandoned driftnets. The agency also evaluated the feasibility of establishing a bounty system to pay persons who retrieve lost or abandoned driftnets and other plastic fishing gear from U.S. waters. A report from the February 1988 fishermen's workshop on this issue (see section on Sea Grant in Chapter V) is due to Congress in June 1989.

These legislative initiatives are bringing the resources of the federal government to bear on the problem of plastic debris. Now that more attention has been focused on the issue, it is expected that federal programs will use their existing authority more forcefully to attack the problem.

Pollution Laws

Another type of measure that has been examined as a way to stop plastic pollution are laws that aim to keep our land and water safe and clean. Again, though, these laws were designed to get at harmful chemicals, bacteria, and hazardous wastes, none of which is descriptive of plastic. The Clean Water Act, for example, is aimed at reducing oxygen-demanding wastes, bacteria, suspended solids, and other materials resulting from domestic and municipal sewage and industrial processes that degrade the quality of water. Under this law plastics are not considered a regulated substance.

Although limited in scope, it is thought that most of the plastic resin pellets that get into the ocean are lost during transportation, not actual manufacture. However, if any are discharged from factory pipes they are subject to control under the Clean Water Act.

The Ocean Dumping Ban Act of 1988 under the Marine Protection, Research and Sanctuaries Act, prohibits ocean dumping of sewage sludge and industrial waste after December 31, 1991. The primary purpose of the Ocean Dumping Ban Act is to prevent unreasonable degradation of the marine environment. The Act also includes provisions prohibiting public vessels from disposing of potentially infectious medical wastes at sea and the Act toughens the penalties for dumping medical wastes in the ocean. Although the Ocean Dumping Ban Act addresses the disposal of wastes that "unreasonably degrade the environment," it does not address the handling of plastics in the ocean.

Litter tossed by people is another land-based source of plastic debris. In an effort to get at this problem, many states have litter laws, and have further targeted plastic trash by banning certain plastic containers and beverage container connectors. Eighteen states have banned non-degradable plastic six-pack connectors. A number of other states and local communities are beginning to ban other plastic products. For example, Suffolk County, New York passed legislation effective in 1989 banning the use of non-degradable plastic grocery bags and certain other plastic products added at the retail level, such as those used by fast-food restaurants and convenience stores.

47

Another approach being considered by some states is mandatory separation of trash. For example, New Jersey has mandated statewide separation of three materials so that these materials can be collected for recycling, rather than taking up scarce space in overloaded landfills. Consumers may choose from paper, glass, plastics or aluminum. Rhode Island, too, has mandatory recycling.

WILDLIFE CONSERVATION LAWS

Another body of law that is applicable to the problem of plastic debris, especially where it is the cause of entanglement of marine animals, is wildlife protection law. Under this broad category are such laws as the Marine Mammal Protection Act (MMPA), the Endangered Species Act, the Migratory Bird Treaty Act and the Fishery Conservation and Management Act (FCMA).

Only one, the FCMA, specifically prohibits the disposal of nets into U.S. waters. This prohibition, however, only applies to foreign fishermen operating in U.S. waters. Prior to passage of MARPOL Annex V and the MPPRCA, there were no regulations prohibiting American fishermen from dumping their unwanted nets overboard.

Under the MMPA, Congress several years ago authorized the first and only program dealing specifically with entanglement of marine wildlife in plastic debris, addressing the problem through research and education. There is a prohibition under the MMPA against "taking," or killing, marine mammals, and under certain circumstances intentional or negligent discard of fishing nets might be prosecuted as an illegal "take" of a marine mammal.

Other wildlife laws prohibit taking certain protected species, and a drowning by entanglement could be considered a "take." The problem with this approach, though, is identifying the source of the entangling debris. As with the ocean dumping laws, unless the discard was actually observed at sea, enforcement is virtually impossible. With increased attention focused on plastic debris as a source of harm to wildlife, endangerment to vessels, and as an expensive and unsightly nuisance on our beaches, new ideas and schemes for regulating disposal of plastics are being considered at the local, state, national, and international levels. Some of the approaches focus on plastic itself, some on the people who use it, and others on the marine animals harmed by it.

Perhaps more important, however, than designing new laws to control how we make or use plastic, is to be informed about the effects of the improper disposal of plastic debris. Informed consumers and industrial users can reduce the harmful effects of their plastic trash by discarding it properly.

V

SO WHAT'S THE SOLUTION?

The preceding chapters describe the pervasive and serious problems caused by improper disposal of plastics in the marine environment, but the picture is not entirely bleak. With growing citizen and government awareness of these issues has come action—action by the industries that contribute to marine debris, action by individuals and private conservation groups, and action by government. Unlike the case with many environmental issues that have too often set industry and conservationists on opposite sides of the fence, the problem of marine debris has created an unprecedented amount of cooperation, information sharing, and jointly sponsored efforts to find solutions.

INDUSTRY ACTION

PLASTICS INDUSTRY

The Society of the Plastics Industry (SPI) is a trade organization of more than 1900 members representing all segments of the plastics industry in the United States, including resin producers, distributors, machinery manufacturers, plastics processors, and moldmakers. In November 1986, SPI held a meeting with representatives of major resin companies to discuss the problem of resin pellet escapement into marine areas, and SPI is currently conducting a comprehensive survey of industry practices concerning resin pellets. Information gathered thus far indicates that most pellet escapement may have resulted from past industry practices that have now been corrected, such as inadequate screening of factory effluent.

Education

SPI, in cooperation with the Center for Marine Conservation and the National Oceanic and Atmospheric Administration, has developed a public education campaign to address improper disposal of plastics in the marine environment. The first phase of this project was a series of public service announcements for trade publications of the merchant shipping, commercial fishing, and plastics industry with accompanying brochures targeted at each group. Similar types of materials have been produced for recreational boaters and sports fishermen. NOAA and SPI also support hands on activities at fishing tournaments, trade conferences and beach cleanups.

Besides education, there are two additional areas where the plastics industry can help: degradability and recycling of plastic products.

Degradability

Because of growing interest in degradable plastics, SPI sponsored the Symposium on Degradable Plastics in 1987, which was attended by over 300 persons. The purpose of the symposium was to disseminate information about work being done within and outside of the plastics industry, and to explore the technical, economic, and social dimensions of this subject. The objectives were to examine not only if and how plastics can be made to degrade but also whether it is desirable for plastic products to be made degradable. In some cases the qualities required for certain products, including safety requirements, could be compromised by steps taken to make them degrade. For instance, one would not want hazardous liquids to be contained in a bottle that could degrade. In general, promising work is being done to make some plastics either biodegradable or photodegradable but degradability should not be viewed as the panacea for solving the marine debris problem.

There are also several questions that remain with respect to degradable plastics. For instance, what are the effects on the environment of the byproducts of degradable plastics? A copy of the proceedings from the symposium, which contains valuable information on existing degradable technology, is available from SPI.

Recycling

"Plastic" is a generic term. In reality there are hundreds of different types of plastic, each with its own special characteristics which make it best suited for different types of products. Although plastic bottles, for instance, may look somewhat similar, they are made from different types of plastics. And different kinds of plastics could not be recycled together. (However, there are some recent developments in mixed-plastics recycling.) Therefore, the first recycling attempts have concentrated on bottles that are made of the same type of plastics and that are easily identifiable by consumers and recyclers.

The concept of recycling is not new. Plastic bottles were introduced in the late 1940s. At that time scrap plastic created during the manufacture of plastic bottles was recycled back into the manufacturing process. By the mid-1970s creative entrepreneurs were recycling plastic milk bottles into toys, pails, and commercial flowerpots. Plastic milk bottles were a logical place to start since they were easily identifiable, they were all made of the same type of plastic (high-density polyethylene (HDPE)) and there were enough milk bottles available to provide a fairly abundant and constant supply for recycling. The introduction of plastic soft drink bottles in 1978 added to the recycling pool. In fact, recycling of plastic soft drink bottles began in 1979, one year after they were introduced to consumers. Soft drink bottles are made of a different type of plastic, polyethylene terephthalate (PET). In 1987, plastic milk jugs and soft drink bottles composed about 40 percent of all plastic bottles produced in the United States. In 1987, 150 million pounds of PET bottles—or about 20% of all those sold—were recycled.

Because of concerns about purity and food contact, recycled beverage containers are turned into other products and not back into the original form. Today one use for recycled plastic milk jugs is the manufacture of plastic "lumber" which does not have to be painted because the color is pigmented in the material. Other items made from recycled milk containers include underground pipes, toys, pails and drums, traffic barrier cones, garden furniture, golf bag liners, kitchen drain boards, milk bottle carriers, trash cans, and signs. Fiberfill is the major end-use for recycled soft drink PET. Fiberfill is used as a stuffing for pillows, ski jackets, sleeping bags, and automobile seats; 36 plastic soft drink bottles will produce enough fiberfill for a sleeping bag; 5 bottles make enough for a man's small ski vest. Plastic strapping is another major use for recycled PET.

The Plastic Bottle Institute, a division of SPI, sponsors a continuing program to promote the recycling of plastic bottles into new products. The Institute also encourages the incineration of combustible solid waste in waste-to-energy systems as a viable means of easing the burden on landfills. The Institute produces several types of informational materials on plastic recycling including an annual Directory and Reference Guide to companies involved in recycling, fact sheets and brochures, and The Plastic Bottle Reporter, a quarterly newsletter.

The Plastic Bottle Institute was instrumental in the formation of the Plastics Recycling Foundation in 1984. The purpose of this nonprofit foundation, funded through industry contributions and government and university grants, is to research and develop improved technology of plastics recycling to make it more practical for businesses nationwide. Once this is accomplished the amount of plastics that can be recycled into useful new products is expected to increase significantly. Along with the beverage industry, the plastics industry has committed millions of dollars to the Plastics Recycling Foundation to advance research to accelerate recycling of PET. The Foundation has established the Center for Plastics Recycling Research at Rutgers University in New Jersey, where the majority of the research and development work takes place.

To add recycling efforts, many states are now adopting a voluntary plastic container coding system. The code system identifies the six most common plastic resins to assist recyclers conduct efficient, profitable operations.

Most recently, to address the nation's growing solid waste problem SPI formed the Council on Plastics and Packaging in the Environment (COPPE). COPPE is a broad-based coalition composed of representatives from the plastics, packaging, food and beverage, convenience restaurant, and related industries. The goal of COPPE is to develop information relevant to the composition and the disposal of solid waste and then disseminate materials to public officials, environmental groups, and the general public. Waste management planners in municipalities across the country are in need of complete and accurate information about the options and solutions available to them.

In 1988, the plastics industry established The Council for Solid Waste Solutions (CSWS) to deal with the U.S. waste management problem. CSWS supports technical research, government relations and communications programs, to work towards the adoption of long-term, environmentally sound disposal and reuse of plastics. For more information on COPPE, CSWS, and these various activities of SPI, see Appendix I.

COMMERCIAL FISHING INDUSTRY

Due primarily to the successful education efforts of the Marine Entanglement Research Program (see "Department of Commerce" below), commercial fishermen on the Pacific Coast have become increasingly aware of the marine debris problem and their contribution to it. But they have also demonstrated an outstanding willingness to help mitigate this problem—in many areas "Stow it—don't throw it" has become the motto of the fishing industry. In late 1987, a coalition of commercial fishermen sponsored the North Pacific Rim Fishermen's Conference on Marine Debris. Approximately 60 representatives from the fishing industries of the United States, Canada, Japan, the Republic of Korea, and the Republic of China (Taiwan) attended this five-day meeting with the goal of reducing the amount of debris originating from commercial fishing vessels operating in the North Pacific Ocean. Topics discussed included the nature and magnitude of the debris problem, the legal framework, actions and programs currently being undertaken by the fishing industry to adress the problem, and technical problems and solutions. On the last day, attendees outlined the following set of goals for the fishing industry to address this problem:

1. Every effort should be made to insure that plastic materials are not discarded at sea and loss of fishing gear must be avoided where possible. This goal should be achieved by incineration of non-toxic combustible materials when feasible, retention of synthetic materials for shoreside recycling or disposal and the development of onboard procedures for handling persistent

plastics.

2. A maximum effort should be made to reduce the quantities of synthetic refuse on board by minimal use of plastics packaging materials and through use of washable dishware and other eating utensils.

3. Special attention should be given to promoting the development of affordable technology and operational procedures which will lead to a reduction in the loss of fishing gear and which will enhance the recovery of fishing gears.

4. Early adoption and enforcement of MARPOL Optional Annex V which prohibits discharge of all plastic materials should be promoted along with needed domestic regulations. (Annex V was ratified by the United States and took effect internationally on December 31, 1988.)

5. Because of the global character of marine debris and the multitude of user groups which contribute to the problem the fishing groups involved in the conference will focus their efforts to encourage other industries contributing to the marine debris problem to become involved in seeking solutions.

6. Fishing groups are encouraged to promote local programs to further the education of fishermen, port authorities, resource managers, other seafarers and the general public regarding the scope, magnitude and consequences of the growing marine debris problem.

7. Fishing vessel operators in the North Pacific will be encouraged to post in plain view a notice to officers and crew that discharge of plastic materials into the oceans is contrary to international law which came into force in December 1988.

8. Participants in the conference will encourage their organizations to cooperate with dock/port authorities and other government agencies to establish effective shoreside refuse disposal systems.

Fishermen in the Atlantic and Gulf of Mexico have already demonstrated a willingness to follow the lead of their Pacific counterparts and are also finding ways to address the plastic debris problem.

PORT OF NEWPORT, OREGON: MODEL PORT

U.S. ratification of Annex V means that all vessels must dispose of their refuse dockside. Therefore, expansion of shore facilities to handle the increased demand will be necessary. But large, heavy, and bulky items are not easily disposed of. The National Marine Fisheries Service confronted this problem in 1987 under a cooperative agreement with the Port of Newport, Oregon, a small port with an active sport and commercial fishing fleet. The goal was to establish and operate a model project for retrieval of vessel-generated refuse. But without the hard work and dedication of Fran Recht, the project would probably never have been as successful as it is today. Ms. Recht is the director of the Marine Refuse Disposal Project at the Port of Newport—a one-year pilot project to encourage commercial fishermen and recreational boaters to bring their trash back to port. The major objectives of this project were to increase awareness within the boating and fishing community of the dangers caused by at-sea disposal of plastics, encourage proper disposal of these materials, and design a model system to receive and dispose of vessel-generated wastes. With the assistance of an advisory committee composed of representatives from the commercial fishing groups, marinas, Coast Guard, and others, Ms. Recht designed a system whereby dumpsters were not only easily accessible but supplemented by bright blue recycling boxes marked for cardboard, metal, cable, wood, and netting.

As a result of these efforts, use of port trash disposal facilities has increased dramatically—fishermen are bringing back more garbage than ever before. But the money spent by the port to dispose of refuse once it has been brought in has actually decreased by 6.5 percent. These savings are due to the fact that fishermen are voluntarily using the port's recycling bins for bulky items like cardboard and nets so that dumpster space for domestic wastes can be fully utilized. The port, however, does not make any money from recycling. Actually, the "recycling" is carried out by the local residents. Cardboard is taken by a local hauler for recycling, scrap metal is collected by a metal dealer, and wood is collected by Wood Share, a program conducted under the Gleaners' organization, which collects and processes burnable and otherwise useable wood to give to needy citizens in the community. As for fishing nets, according to Ms. Recht, there's never any problem getting rid of them. People use them for horticulture, childproof barriers around porches, decoration, and even for volleyball nets and backstops for baseball diamonds.

Information on the Marine Refuse Disposal Project is available to other ports interested in establishing their own refuse systems from the Port of Newport (see Appendix I) or the National Marine Fisheries Service Marine Entanglement Research Program (see Appendix B).

DON'T TEACH YOUR TRASH
TO SWIM!

Courtesy of the Marine Refuse Disposal Project, Port of Newport, Oregon

PETROLEUM INDUSTRY

The Offshore Operators Committee (OOC) is an organization of over 60 companies that conduct essentially all the gas and oil exploration and production activities in the Gulf of Mexico and Atlantic Ocean. The OOC was organized in part to promote protection of the marine environment. Although many offshore operators include proper waste disposal practices in employee orientation programs and other educational efforts, debris from this industry is still prevalent in the Gulf. Therefore, in 1985 the OOC produced an educational video on the debris problem. The film focuses on three fictional offshore workers who are deliberately or accidentally responsible for causing litter to enter Gulf waters. These employees later encounter the same type of litter on the beach during their days off. The film is based largely on the premise that some offshore workers may think their litter is out of sight and mind just because they are far from land.

To enhance the value of the movie one of OOC's member companies has produced a hardhat decal with the slogan "Clean Rigs-Clean Water-Clean Beaches." Hardhat decals are a popular way to convey messages in the oilfield.

The OOC's anti-litter efforts were recently recognized by the National Park Service's Take Pride in America Campaign which is a national public awareness campaign to encourage development of a feeling of stewardship of our public lands. Copies of the OOC video and further information on their educational efforts are available from the Committee (see Appendix I).

MERCHANT SHIPPING INDUSTRY

The American Institute of Merchant Shipping (AIMS) is a national trade association representing a large portion of U.S. flag merchant vessels. AIMS supports Annex V of MARPOL as the best means to reduce improper ocean disposal of plastics, and has so stated in many Congressional hearings. The organization supports required pollution prevention training for merchant seamen, including required training on how to achieve compliance with Annex V standards. The organization also supports logbook entries and waste management plans to ensure vessel compliance with proper disposal under Annex V.

PORT AUTHORITIES

Organized in 1912, the American Association of Port Authorities (AAPA) represents virtually all of the public port authorities in the United States and major port agencies in Canada, Latin America, and the Caribbean nations. Generally, public Port Authorities are created by state or local governments to facilitate international trade and to stimulate economic development in the regions they serve.

AAPA surveyed its U.S. members in 1987 to assess the existing availability of disposal facilities for ship garbage near ports. The survey specifically asked about the availability of facilities for disposing of garbage regulated by the U.S. Department of Agriculture (USDA). At that time, the results of the survey indicated that approximately 60 percent of the 84 member ports had facilities that were capable of handling regulated wastes. However, some of them were only capable of handling wastes in emergency situations and not on a routine basis. Among the 40 percent that did not have USDA approved facilities were some of the larger port areas in the country including Seattle, Houston, northern New Jersey and the Delaware River ports. The majority without facilities were said to be smaller ports with limited financial resources. All ports are now seeking ways to comply with Annex V.

STOW IT
DON'T THROW IT!

Center for
Marine Conservation
Formerly Center for Environmental Education. Est. 1972

*"Don't be a
Litter Boat"*

FEDERAL GOVERNMENT ACTION

DEPARTMENT OF COMMERCE

National Oceanic and Atmospheric Administration (NOAA)
National Marine Fisheries Service

In 1984, Congress recognized the problems of marine debris and entanglement, and directed the National Marine Fisheries Service (NMFS) to develop a research program in consultation and concurrence with the Marine Mammal Commission. The goals and objectives of this program, the Marine Entanglement Research Program, were developed from recommendations that came out of the international Workshop on the Fate and Impact of Marine Debris held in Hawaii in 1984.

The Marine Entanglement Research Program (MERP) conducts educational activities aimed at debris generators; oversees the operation of two marine debris information offices; sponsors research on the origin, amount, distribution and fate of marine debris; and explores ways to reduce the amount of non-degradable material lost or disposed of at sea. A detailed list of MERP education, research, and mitigation projects, as well as a list of reports available from MERP, can be found in Appendix B.

In addition to its own research efforts, the Marine Entanglement Research Program has brought together experts from the plastics manufacturing, merchant shipping, commercial fishing, and solid waste management industries, as well as government representatives and members of the conservation community to share their perspectives on problems and solutions. This advisory group, called the Marine Debris Roundtable, has encouraged discussion and advancement of ideas and technology from degradability to waste management plans. The exchange of information among these volunteer representatives has resulted in initiatives by the private sector as well as encouragement to other federal agencies.

National Marine Pollution Program Office

The National Ocean Pollution Planning Act of 1978 calls for the establishment of a comprehensive, coordinated, and effective federal program for ocean pollution research, development, and monitoring. As required by the Act, NOAA, in consultation with other agencies, prepares a five-year Federal Plan for the National Marine Pollution Program every three years. The National Marine Pollution Program Office (NMPPO) within NOAA updates the five-year Plan and coordinates the implementation of recommendations in the Plan.

In 1987, NMPPO convened a workshop to set national priorities for research on the five most important marine pollution problems. "Persistent marine debris" emerged as one of the top five.

A group of 30 individuals representing the plastics industry, federal and state governments, the commercial and sport fishing industries, merchant shipping, the Navy, academia, and environmental organizations identified these research priorities:

- methods for handling vessel wastes;

- the effects of plastic pellets, particles, and fragments in the marine environment;

- identification of the sources of land-based litter;

- identification of the sources of plastic resin pellets;

- investigation into effects of "ghost fishing" traps; and

- examination of methods to alter, modify, substitute, or regulate fishing nets, traps, rope, and line to reduce their harmful effects as marine debris.

The results of this workshop were presented in proceedings entitled "National Marine Pollution Problems and Needs Workshop," June 9–11, 1987 in Easton, MD. This document was the framework for the congressional report outlining the "Federal Plan for Ocean Pollution Research, Development, and Monitoring Fiscal Years 1988–92."

Sea Grant College Program

The National Sea Grant College Program has recently stepped up its efforts to address the marine debris problem. Sea Grant's marine advisory agents and communicators are involved in educational efforts on the local and regional level. Several Sea Grant offices in various states have taken on the responsibility of organizing citizen beach cleanups during COASTWEEKS (see Appendix C for list of Sea Grant Offices).

In February 1988, the University of Alaska Sea Grant College Program with the assistance of Sea Grant programs in several other states sponsored a workshop to discuss incentives for fishermen that would encourage them to bring back old gear and to retrieve lost gear they find in the marine environment. The workshop was held in response to the recent passage of legislation that requires NOAA to specifically evaluate the feasibility of establishing a driftnet marking, registry, and identification system and a bounty system to pay persons who retrieve lost or abandoned driftnets and other plastic fishing gear from U.S. waters.

Topics discussed included technical approaches, such as gear marking systems and degradable plastics applications, as well as incentives such as bounty systems for the retrieval of lost gear and a deposit-refund system for gear purchases (similar to beverage bottle deposit systems). However, fishermen attending the meeting strongly suggested that education efforts directed at fishermen would be the best way to enlist this industry's cooperation.

MARINE MAMMAL COMMISSION

The Marine Mammal Commission was established in 1974 under Title II of the Marine Mammal Protection Act of 1972. It is responsible for developing, reviewing, and making recommendations on actions and policies for all federal agencies with respect to marine mammal protection and conservation and for carrying out a research program.

The Commission asked the National Marine Fisheries Service to convene the above mentioned 1984 Workshop on the Fate and Impact of Marine Debris. The Commission also provided impetus to the National Marine Pollution Program Office of NOAA to incorporate the marine debris problem into the 1985–1989 five-year National Marine Pollution Program Plan. Together with the Marine Pollution Program Office, the Commission helped fund a study to compile information on the sources, fates, and effects of marine debris in the northwestern Atlantic, the North Sea, the Gulf of Mexico, and coastal waters along the west coast of Baja California and adjacent islands. In 1985, the Commission provided funding to facilitate the organization of beach cleanups and cooperated with the Coast Guard on efforts to assist in the U.S. ratification of MARPOL Annex V.

DEPARTMENT OF TRANSPORTATION

United States Coast Guard

The U.S. Coast Guard is probably the most familiar federal agency to those who live, work, and play on our coasts. It is the enforcement arm of the government when it comes to laws that control discharges of pollutants by vessels, and often leads the response teams that clean up marine spills when they occur. The Coast Guard plays an active role in representing U.S. interests in international meetings aimed at marine pollution prevention and cleanup.

Citizens can help the Coast Guard do its job by reporting observations of vessels that are polluting waters to the nearest United States Coast Guard office. Citizens also can encourage local Coast Guard Auxiliaries, which conduct boater safety courses, to include information on proper trash disposal in their public education programs.

DEPARTMENT OF STATE

The Department of State took the lead role in urging U.S. ratification of MARPOL Annex V. It, too, represents U.S. interests at international meetings on marine pollution issues, and will be instrumental in forging new agreements. For example, the State Department was involved in discussions among nations that border the Caribbean to consider measures that would give more protection from marine pollution to certain areas. The State Department also serves as the principal negotiator for the United States during meetings that set up fisheries agreements with foreign nations that fish in U.S. waters. During these kinds of meetings, expressions of citizen concerns could influence decisions on marine debris issues—for example, provisions that regulate the use of driftnets.

DEPARTMENT OF AGRICULTURE

The Department of Agriculture's (USDA) Animal and Plant Health Inspection Service (APHIS) regulates incoming garbage from ships to prevent the entry of exotic agricultural pests and diseases into the United States. Ship garbage has been found to be the source of diseases such as swine fever and hoof and mouth disease. APHIS regulations on disposal of ship's garbage come into play if vessels docking at U.S. ports have visited a foreign country other than Canada. According to APHIS, any part of food or plants or any type of packaging that comes into contact with food is garbage, and regulations require that garbage be kept in a tightly covered leakproof container while within the territorial waters of the United States or that it be taken off the ship and disposed of on land under the direction and supervision of the USDA. But disposal does not mean simply using a nearby trash can. Garbage that comes into the United States from a ship of foreign origin must be either incinerated, sterilized, or ground up and discharged into an approved sewage system at a USDA approved facility. Ships that travel between U.S. ports, such as from New York to Baltimore, or between continental U.S. ports and Canadian ports are not required to meet APHIS regulations.

ENVIRONMENTAL PROTECTION AGENCY

The Environmental Protection Agency (EPA) has several avenues through which it can approach the marine debris problem. The EPA regulates ocean dumping by issuing permits, and can designate special areas for certain kinds of dumping. It also controls inland sources through its programs that permit industrial discharges into rivers and streams. The EPA is also the agency that sets the standards for sewage treatment and sludge disposal, two more sources of persistent marine debris. Recently, the agency has begun taking a regional approach to coastal and ocean pollution through its "Gulf Initiative." This program is intended to bring together all the federal, state, and local authorities in the Gulf of Mexico region that have control over one or more parts of the ocean pollution problem, and try to view the problems, regulations, and possible solutions on an interdisciplinary, regionwide basis. In 1989, the EPA supports the National Marine Debris Data Base and regional studies of plastic debris in U.S. harbors.

DEPARTMENT OF THE INTERIOR

Minerals Management Service

The U.S. Minerals Management Service (MMS) is the lead regulatory agency over the offshore oil and gas activities in federal waters, and hence the regulations for handling and treating garbage generated during the course of these activities have been established by MMS. These regulatory requirements are set out in a pollution prevention and control order that restricts the disposal of solid waste materials into the ocean. Personnel who oversee these activities conduct routine inspections of offshore operations to determine if such laws are followed. In addition, in 1986 MMS issued a special directive to all Gulf of Mexico lessees and operators advising them to develop and use training aids and awareness programs specifically targeted on the marine debris problem. The MMS also sponsors an annual Information Transfer Meeting for the public, to present major Gulf of Mexico environmental issues and relevant activities. In recent years, the Annual Information Transfer Meeting has held special sessions that address the issue of marine debris.

DEPARTMENT OF DEFENSE

United States Navy

The U.S. Navy has had a Shipboard Pollution Abatement Program since 1971, and has had regulations in place prohibiting the discharge of unpulped solid waste within 25 miles of shore. Navy regulations also require shipboard personnel to attempt to package all vessel waste so that it sinks. The Environmental Protection and Occupational Safety and Health Division of the Navy has been conducting research for some time on various technological alternatives to waste disposal including shredders, pulpers, compactors, and incinerators. Although public vessels, including navy ships, are exempted from the requirements of Annex V, the U.S. Marine Plastic Pollution Research and Control Act of 1987 requires all public vessels to come into compliance with Annex V by 1994. In October 1987, the Navy convened an Ad Hoc Advisory Committee on Plastics, composed of approximately 20 individuals including Congressional staff and several environmental representatives. The final product of this Committee was a document outlining recommended options and components of a comprehensive plan to minimize plastic wastes that the Navy can consider including improved shipboard waste handling facilities, modifications in supply and provisioning to minimize the amount of plastics taken onboard, changes of shipboard procedures and the implementation of education programs on the plastic debris problem for crew members and other naval personnel.

Since the release of the report, the Navy has investigated several methods for shipboard compliance with at-sea discharge of plastics. The Navy hopes to develop new technology including a plastic waste processor that will reduce the volume of plastic trash stored on board 30 times. Most naval ships now have two separate trash cans in all spaces where there used to be one: one specifically for plastics and the second for all other trash. The Navy also sent education packets to all 600 ships in the fleet. The package contained the first edition of this book, posters, a *Ship's Guide to Shipboard Solid and Plastics Waste Management,* and a video featuring Huey Lewis's popular song "(Living in a) Perfect World."

INTERNATIONAL ACTION

INTERNATIONAL MARITIME ORGANIZATION

Established in 1958, the International Maritime Organization (IMO) (formerly the IMCO or International Maritime Consultative Organization) is a specialized agency of the United Nations that deals with international shipping issues, including safety and pollution. As an international organization, the IMCO sponsored conferences that encouraged shipping nations to draw up treaties addressing marine safety issues. Under IMCO auspices, several conferences were held, resulting in international agreements on the control of oil pollution from ships. These conventions eventually led to the adoption in 1973 of MARPOL, officially known as the International Convention for the Prevention of Pollution from Ships.

In 1978, the IMCO sponsored another international conference which produced the MARPOL Protocol of 1978—a measure designed to speed up MARPOL's adoption by the shipping nations of the world. This important protocol expanded the substances to be regulated, addressing not only oil

Top Secret Trash

The fact that U.S. naval vessels routinely discharge trash over the rail may seem deplorable to some. But there are those that look upon this practice as a great opportunity. In September 1986, the aircraft carrier *USS Constellation* was on routine maneuvers in the Gulf of Alaska. When it was time to take out the trash, as usual, it was tossed overboard. But trailing the *Constellation* was a Soviet surveillance ship that quickly came up behind the vessel to pick up the bags. Why? To look for U.S. military secrets.

According to the Senior Intelligence Officer of the *Constellation* the ship was "in typical fashion, getting rid of (its) garbage." According to naval regulations shipboard garbage is put in plastic bags, punched with holes, and then tossed overboard. But apparently one bag didn't sink and the Soviet ship deployed a small vessel to snatch it up.

While all classified information on a naval vessel is destroyed by shredders before it is discarded, there are certain kinds of unclassified information that the Soviets might find by garbage collecting—such as an old aircraft carrier telephone directory. When asked what was in the floating garbage bag that was collected that day, the senior intelligence officer responded, "food, beverage cans, and the garbage of the 5,000 men on board."

---

pollution from ships, but also disposal of other pollutants. The MARPOL Protocol contains five categories, or Annexes, that state the regulations governing specific types of pollution. Annexes I–IV address the prevention of pollution from oil, chemicals, hazardous substances in packaged form, and ship generated sewage. Annex V contains regulations specifically dealing with garbage, including the intentional discard of fishing gear, packing materials, dunnage, and food wastes. A key feature of Annex V is its prohibition on "the disposal of all plastics, including but not limited to synthetic ropes, synthetic fishing nets, and garbage bags."

UNITED NATIONS ENVIRONMENT PROGRAMME/ REGIONAL SEAS PROGRAMME

The Regional Seas Programme is designed to link the status of a region's marine environment with actions that can combat environmental deterioration. The program does this by helping nations bordering common seas, such as the Mediterranean, reach regional agreements for cooperation in addressing marine pollution problems. There are presently eleven regional seas programs in various stages, ranging from research to action plans to completed treaties. One example is the Convention for the Protection and Development of the Marine Environment of the Wider Caribbean Region. This international agreement, to which the United States is a party, sets out a process by which countries bordering the Caribbean can work cooperatively to solve problems like oil pollution, land-based sources of pollution and other issues. Parties may soon be considering ways this agreement can be used to target the persistent debris problem in the Caribbean, particularly the Gulf of Mexico.

UNITED NATIONS FOOD AND AGRICULTURE ORGANIZATION (FAO)

The fisheries division of the FAO has been collecting information on the entanglement of marine resources in fishing nets and debris, and also has cooperated with many other UN organizations to gather and coordinate data on marine debris and pollution issues.

OTHER INTERNATIONAL EFFORTS

Not only the United Nations, but many other organizations such as the Commission of the European Economic Communities, the Intergovernmental Oceanic Commission, the International Union for the Conservation of Nature and Natural Resources, and countless other academic, professional and governmental bodies have research projects underway or are considering proposed studies that will target the occurrence and impacts of persistent marine debris in our oceans.

One of the major frustrations facing policy analysts and decision makers is that none of this work is coordinated nor is the data correlated. This makes it difficult to get a global picture of the persistent debris problem and to propose cooperative solutions. However, it is hoped that with increasing public awareness of the issue, sharing of information among researchers, and increased efforts by governments to control improper disposal of harmful debris, we will begin to get a handle on this worldwide problem. The Second International Conference on Marine Debris, held in April 1989, brought together scientists and experts from around the world, and facilitated this exchange of information.

VI

GETTING INTO THE ACTION: THE CITIZEN'S ROLE

I n the previous chapters we have seen that the problems caused by marine debris stem from a broad range of sources both on land and at sea. But since each of us uses plastics in one form or another, each one of us also has the opportunity every day to become part of the solution to the problems.

First, public awareness about the problems caused by marine debris is essential not only at the individual level but at the community and larger political level to inspire national and international cooperative efforts to solve these problems. Much can be done at the community level to promote awareness of the need to dispose of plastics properly. State beach cleanups and local recycling projects are good examples. Because the issue involves a diversity of interests from industry to concerned citizens, it is likely that different aspects of the problem will be of concern to different groups. For instance, fishermen may be concerned about the effects of ghost fishing; the local tourist bureau may be more concerned with aesthetic degradation of marine areas. Each group will want to focus on the aspects of the problem with which it is most directly concerned. But the ultimate goal is the same— to keep plastics out of the ocean.

SET AN EXAMPLE

Just by reading this guide you have taken the first step toward learning more about the problem. Now practice and promote proper disposal of plastics in your home, at the beach, and if you own or operate a boat, in lakes, rivers, and at sea.

At home, be sure not to dispose of plastics in sewer systems (e.g. plastic tampon applicators). At the beach, be sure that plastic wastes are disposed of in trash facilities. If these facilities are inadequate tell the person or agency responsible for maintaining the beaches. Remember to break or cut the loops of plastic six-pack holders before disposing of them to ensure that if the ring escapes into the water it will not entangle an animal. And look for degradable holders when making your purchases (see Chapter IV for a list of states that require degradable six-pack connectors). In addition, fishermen should take special care in properly disposing of lengths of old fishing line since lost line can be lethal.

On a boat, be sure to stow your plastic trash and old fishing gear for proper disposal on land. If you are the captain make it ship policy. In addition, consider using reusable items such as washable dinnerware to minimize the amount of plastic waste you generate. One way to manage your trash is to secure trash bags and cans to a fixture on the boat and then make sure that all trash finds its way into this receptacle. Then when you come into a marina or dock you can dispose of it properly.

LINDA MARANISS:
THE FIRST TEXAS BEACH BUDDY

When Linda Maraniss opened a regional office of the Center for Marine Conservation (formerly the Center for Environmental Education) in Texas she expected to spend most of her time working to protect endangered sea turtles from drowning in shrimp nets. But a pleasure trip to the beach changed her plans. Linda was a newcomer to Texas, and was eager to explore all the state had to offer. What the beaches offered her was trash. Tons of it, dangerous and ugly.

She was so disturbed that she decided her first priority as director of CMC's regional office was cleaning up the Texas coastline. She was familiar with Judie Neilson's successful beach cleanups in Oregon (p. 11) and wanted to base the Texas cleanup on that model.

She made hundreds of phone calls and sent hundreds of letters to garner volunteers, endorsements, money, and supplies. Her theme was "Be a Beach Buddy" and it quickly caught on as posters and brochures were distributed throughout the state, encouraging people to attend the cleanup on September 20, 1986. She enlisted volunteer "zone captains" to coordinate the activities at each location, and local businesses were encouraged to lend their support. Radio stations sponsored songwriting contests, and hotels offered discounts to cleanup volunteers. The result? Nearly 3,000 people rid 122 miles of beach of 125 tons of debris.

From the start, Linda knew she wanted the cleanup to be more than a three-hour litter pickup. Volunteers filled out data cards on the types and amounts of debris they found (see Appendix M). Linda put together a steering committee of experts to analyze the data and make recommendations for solutions. The final report on the cleanup contained no less than 29 recommendations to reduce marine debris.

The cleanup generated a number of new state regulations and programs, such as the Adopt-a-Beach program (see p. 73). Beach buddies, whose eyes had been opened to the debris problem, looked forward to future cleanups, saying "See you again next year!" as they departed. Federal agencies would later hold onsite inspections of Texas beaches to learn more about the debris problem.

Linda had raised the public consciousness and there was no turning back. By 1987 the number of beach buddies had swelled to more than 7,000 and cleanup sites extended from the Mexican border to the Louisiana state line. For copies of the 1987 report contact the Marine Debris Information Office nearest you. (See Appendix B.)

If facilities on land are inadequate, express your concerns to the marina owner, dock master, or port authorities. If enough people express their concern the facilities will most probably be upgraded. If not, tell the Coast Guard. They are responsible for keeping navigable waters litter-free. Also notify the Coast Guard if you see other boat crews dumping trash overboard. Be sure to get the vessel's name, number, location, and type of trash. State and local officials should also be informed of inadequate waste facilities in marine areas.

COMMISSIONER GARRY MAURO'S TEXAS ADOPT-A-BEACH PROGRAM

On September 20th, 1986 the Center for Marine Conservation's Gulf Coast States Regional Office sponsored the first statewide beach cleanup in the state of Texas. Orchestrated by director Linda Maraniss, the cleanup was a tremendous success—approximately 3,000 volunteers came out to the beaches and spent three hours picking up trash. But for one "Beach Buddy" who participated that day the cleanup would go on to be more than just a one day event.

Maraniss had persuaded Texas State Land Commissioner Garry Mauro to help in the cleanup. Mauro says he expected to go out to the beach, pick up a few cans and bottles, and go home. But the volume of trash collected that day opened his eyes to the extent of the debris problem in Texas. Since that day Commissioner Mauro and his staff have gone on to organize an Adopt-A-Beach program in Texas where groups or individuals adopt a particular section of beach for one year and sponsor at least three cleanup efforts at the site.

By 1988, all of the 172 miles of accessible coastline in Texas has been adopted and is being cared for by concerned citizens. Commissioner Mauro has become a leader in the campaign to fight marine debris.

INVESTIGATE THE PROBLEM IN YOUR AREA

What types of marine debris do you notice most frequently? The types of debris can often point to the sources of the debris. Are fishing gear, strapping bands, or plastic sewage-associated wastes prevalent in your area? Are there fishing fleets, merchant vessels, marinas, or industries in your area that could be a source of this debris?

A more detailed analysis of debris can be conducted using the data card provided at the end of this guide. This data card was developed by the Center for Marine Conservation specifically for volunteers participating in beach cleanups along the Gulf of Mexico and is now used on beach cleanups nationwide. See Appendix M and insert titled "47,000 Citizens" for more information on data cards and the National Marine Debris Data Base.

After recording data on the types of debris found on a section of beach, count the number of plastic items in relation to those materials made of glass, metal, paper, and rubber. Most likely plastics will compose the greatest number of items. To determine the sources keep in mind that while most plastic soda bottles and bags are anonymous and untraceable, their occurrence with items such as fishing gear, strapping bands, and hardhats may suggest that a portion of the untraceable items may also come from the same sources.

If you have identified ocean sources such as commercial fishing and merchant ships, examine the trash facilities at docks, marinas, and ports in the area. Could the problem in your area be reduced by improving these facilities, thereby encouraging vessels to properly dispose of wastes?

If plastic litter indicates that land-based sources are contributing to the problem, such as plastics industries, sewage systems, and landfills, inquire if any procedures to prevent plastic escapement are carried out at any local industries, treatment plants, or landfills.

In the case of both ocean and land-based sources of debris, if it appears that facilities and mechanisms to prevent marine debris are not being utilized, try to meet with representatives of these groups, or public officials in charge of facilities, to discuss the marine debris problem and possible improvements.

INFORM OTHERS OF THE PROBLEM

Talk with children in your area about the problems caused by the improper disposal of plastic wastes. Encourage local schools to include the topic in their curricula. They may even be interested in field trips to aquatic areas to carry out a beach cleanup using data cards. Zoos, parks, libraries, research centers, and other areas of high community visibility should have information available

MARK SCHRADER: WORLD SAILOR

Asking a sailor not to throw his plastic garbage overboard seems like a simple request. After all, they have seen more than enough ocean debris to understand the problem. But for someone sailing around the world alone—facing collisions with merchant ships and icebergs, falling overboard during storms, and even suffering from hallucinations during unbearable periods of calm—being environmentally conscientious may not be a priority. But for one special sailor it was.

Mark Schrader was one of 24 participants in the 1986 BOC Challenge, a single-handed sailboat race around the world. Like most sailors, Schrader routinely dumped his trash overboard during long voyages. But after meeting with staff from the Center for Marine Conservation and learning about the problems caused by plastics in the oceans he made a commitment to keep all plastic trash onboard until he could dispose of it properly in four ports of call. He went on to enlist the support of all the other BOC skippers. "Our adventure is built on the willingness to live in harmony with the sea. How can we uphold this and use the same sea as a dumpsite for our wastes?" he asked them.

Nine months and 27,000 miles later Schrader returned to Newport, Rhode Island where the race had begun on August 30, 1986. He told of seeing an albatross flying above his boat many miles off Cape Horn. Entangled and trailing from one of its wings was a piece of fishing line about twelve feet long.

On the last leg of his voyage from Rio de Janeiro to Newport, after traveling for 30 days and over 5,016 nautical miles, he saved 438 pieces of plastic, including 71 plastic wrappers, 69 plastic bags, 46 plastic straws, 39 plastic caps and lids, 30 plastic bottles, and 183 miscellaneous plastic items including one piece of electrical tape, and ten candy wrappers.

Schrader will be the Race Chairman of the next BOC Challenge in 1990–1991 and plans to continue his efforts to encourage all sailors to save plastic trash for proper disposal on land.

to the public. If they don't, tell them about the problem and encourage them to write to the groups listed in the Appendices for information. Under contract to NOAA the Center for Marine Conservation operates two Marine Debris

Information Offices where citizens can obtain educational materials, brochures, photographs, and other information on marine debris and wildlife entanglement. (See Appendix B for addresses.)

Publicize the problems caused by plastics in a local newspaper or community newsletter. If you are aware of projects underway in your community publicize them as well.

BECOME PART OF A LARGER GROUP

Look for other individuals or organizations in your community that have an interest in this problem and may already be working toward and implementing solutions. For example, the National Marine Fisheries Service and the U.S. Fish and Wildlife Service have granted permits to a number of individuals and organizations who voluntarily respond to marine mammal and sea turtle strandings. In fact, in every coastal state there is at least one group that responds to marine mammal, sea turtle, and in some cases seabird strandings, including animals that have become entangled in debris. A listing of these organizations is given in Appendix G. To assist in their efforts you could make sure that their phone number is posted in marine areas. In many cases they are in need of volunteers to respond to strandings. Even if you find an animal that has died report it to your local stranding network. Any and all information is important.

But don't look just for those groups who have a common interest with you. If you are concerned about the effects of plastic trash on marine wildlife you may want to work with people who are developing port facilities in compliance with Annex V or fishermen who retrieve lost nets to prevent ghost fishing. Your objectives may be slightly different, but the results will be the same.

And if you know of individuals or groups who oppose your efforts meet with them to discuss the problem. They may not fully understand the problem or your activities. Discuss how you can work together. They may be able to provide useful information or other assistance. When conservationists met with representatives of the plastics industry to discuss the problems caused by plastic pellets two things were accomplished: industry learned of the problem of plastic pellet ingestion, and both groups initiated a joint campaign to inform the public of the problems caused by improper disposal of plastics.

Groups within your state may already be involved in some aspects of the marine debris problem. For instance, many state organizations sponsor activities to increase public awareness about the marine environment and debris during COASTWEEKS, held every year in early fall. A list of orga-

BRUCE BINGHAM: CITIZEN ACTION AT SEA

Citizen action can take many forms. Here is an example of how one concerned citizen, Mr. Bruce Bingham, a naval architect and marine engineer, decided to confront marine debris.

The SS *Veracruz* was returning to its home port of Tampa after a day and a half cruise when the serenity of the evening was broken by several loud splashes. Bruce, his wife Susan, and a number of other passengers were witnessing 40 minutes of bag after bag of garbage being jettisoned from the ship's side. "Looking aft, illuminated by the ship's stern light, the passengers could see a trail of junk disappearing over the waves into the darkness," said Mr. Bingham.

Mr. Bingham tried to work with the ship's captain, Bermuda Star Line manager of operations, and the Tampa port captain to stop the ship from using the ocean as its garbage can. He encountered many obstacles in his inquiries but he eventually found that the boat was legally entitled to dump its garbage overboard. "All ships do it," he was told.

At the time of Mr. Bingham's cruise, in 1988, dumping plastic trash from cruise ships was not illegal. Although he realized that dumping plastics from ships would become illegal in 1989, Mr. Bingham hoped to convince the ship that stopping its dumping practices before the deadline would result in positive public relations in addition to keeping the waterways clear and enhancing the ecology of the area.

Persistence and perserverance payed off. Mr. Bingham won his battle with the Bermuda Star Line. In July of 1988, the company began investigating and making recommendations on ways to stop dumping garbage at sea. On July 30, 1988, the cruise line told its employees that use of the bay as a dumping ground was to be stopped.

Fortunately, as of December 31, 1988, it became illegal for all ships, including cruise liners, to dump plastics at sea. But it's a big ocean, and difficult to patrol all the time. Therefore, citizen action, such as that taken by Mr. Bingham, will be valuable in enforcing Annex V and protecting marine resources. Anyone citing an illegal discharge should report it to the local captain of the port with the name of the vessel, the location of the vessel, the distance offshore, the type and amount of garbage, and provide pictures as evidence, if possible. If the observation leads to a conviction, the citizen could receive a portion of the penalty as a reward.

77

nizations that participated in recent COASTWEEKS activities is given in Appendix E.

47,000 Citizens: National Marine Debris Data Base

In 1988, over 47,000 citizens from 25 coastal states and U.S. territories made an educated guess about the types and quantities of debris found on our beaches. Participants of COASTWEEKS '88 recorded this information on CMC data cards (see Appendix M).

This data provides some very useful information. In all 25 states, plastics represent 55–65 percent of all debris items collected which is consistent with more rigorous scientific beach surveys. On the other hand, medical debris, which received a tremendous amount of press attention preceding the clean-ups, accounted for only one-tenth of one percent of all the debris items. Most importantly, the data cards increase the educational value of a cleanup.

As CMC refines and increases the size of the data base, more will be revealed about the sources of and solutions to marine debris. 1988 was the last year before the MARPOL Annex V plastic ocean dumping ban took effect, and so 1988 data will be the baseline of information for all future data collection efforts. In 50 years analysts will compare information from your grandchildren's data cards with the data card you complete this year. So, remember to make clear marks on the card when you pick up your 33rd plastic milk bottle!

Your state department of environmental conservation, natural resources, or fish and game may have other programs which use volunteers or publicize the importance of being a careful user of the environment. Contact them and find out how you can help. For example, does your area have annual sport fishing tournaments? Why not contact the organizers and volunteer to promote a "Stow it—don't throw it" idea at the tournament? You can offer to distribute information about the plastic debris problem at the tournament and ask fishermen to store plastic trash onboard their boats for proper disposal on land. Make the event fun by holding a raffle for a prize to all fishermen who bring back their trash.

Encourage your state agencies to take advantage of some of the many resources, materials, and information offered by the groups listed in the Appendices.

State agencies can also:

- Provide assistance to beach cleanups in the form of garbage trucks, trash bags, advertising in newsletters and magazines.

- Assess current state laws regarding littering in marine and freshwater areas and the record of enforcement.

- Improve public dock and marina waste disposal facilities.

- Issue information about plastic debris along with state boater registration information, and commercial and recreational fishing licenses.

- Determine the feasibility of plastics recycling and use of degradable plastics in the state.

VII

THE FUTURE:
PUTTING PLASTIC
DEBRIS IN ITS PLACE

Plastic trash in the ocean poses a growing threat for marine wildlife, and a problem for communities and user groups who depend on the ocean. Although we don't know the total amount of plastic trash in the ocean nor the total number of marine wildlife deaths caused by plastic debris, we do know that the persistent nature of plastics means that if we ignore the problem it can only get worse.

Recent congressional action calls for agency studies to identify and quantify the harmful effects of discarding plastics in the marine environment. Assessments of land-based disposal practices, feasibility studies on degradability or substitute materials for the most harmful plastic items, evaluation of recycling and incentives for proper waste handling, and examination of the potential for marking and tracking plastic fishing gear must all be undertaken and completed in the near future. The results of these studies may reveal presently untried solutions.

Despite the importance of learning more about the problems caused by improperly disposed plastic, however, it is clear that many of the effects of marine debris are known, and that more time to study effects should not be used as an excuse to delay needed action.

Ideally, ratification of Annex V will control the amount of plastic trash now entering the oceans from merchant, fishing, passenger, and recreational vessels at sea. But much more than the force of law will be needed for the Annex to be effective. After all, while the ocean is not big enough to handle the incoming tide of plastics, it is large enough to hide offenders who will continue to dispose of plastics at sea. So, seafarers must be encouraged to dispose of plastic trash at shore-based facilities and such facilities must be convenient and easily accessible. Wherever possible, incentives should be provided to encourage the proper disposal of plastics. Above all, crew members must be made aware of the basis and importance of laws that prohibit the disposal of plastics at sea—it's much more than a litter problem.

On the other hand, Annex V regulates but does not prohibit ocean disposal of non-plastic debris. In some areas, materials other than plastic can be a problem, whether they come from land-based or ocean-based sources. This is the case where tidal action and coastal configuration regularly deposit ocean-borne debris ashore, causing severe beach degradation and accompanying economic loss. The Gulf of Mexico is one of those areas where not only plastic, but wood, metal, rubber, paper, glass, and cloth also can be costly, dangerous, and unsightly additions to beaches and estuaries.

One portion of Annex V regulations is aimed at reducing this type of ship's waste in particular ocean regions, called "special areas." According to MARPOL's regulations, special areas are those which for oceanographic and environmental reasons are more susceptible to pollution by garbage. A designation as a special area provides extra environmental protection by prohibiting the dumping of materials in addition to plastic. Such a designation in the Gulf of Mexico would mean that ships could not dump any garbage except ground food wastes into nearshore Gulf waters.

In November 1987, U.S. officials on the Marine Environment Protection Committee of the International Maritime Organization (IMO) informed the IMO that they would seek designation of the Gulf of Mexico as a special area under Annex V. Whether the Gulf of Mexico is designated as a special area will depend on scientific criteria, shipping traffic patterns, and on the reactions of affected user groups, such as the shipping industry, port authorities, commercial fishermen and interested citizens.

Neither Annex V as it exists now, nor the potential of declaring the Gulf of Mexico a Special Area, however, will focus on the remaining crucial component of the marine debris problem: land-based sources. While plastic debris from land-based sources is identifiable by its type (domestic containers, personal products, medical wastes, etc.), no one has yet been able to target how much there is, where it comes from, or how to stop it. This is a subject of appropriate inquiry, particularly now after the Islip, New York garbage barge in 1987 and closed beaches in New Jersey have made their impression on the national psyche: we are no longer capable of handling our overflowing wastes and the ocean is bearing the brunt of that failure.

Congress has mandated several studies of land-based debris, including a three-year, $3 million program to study garbage, sewage sludge, and other forms of pollution in the New York Bight and come up with a restoration plan to clean up that area. In another effort, the EPA has undertaken an integrated, region-wide approach to assessing and controlling all sources of pollutants in the Gulf of Mexico, called the Gulf Initiative. It is hoped that by drawing upon the expertise and resources at municipal, state, and federal levels, officials can come up with some region-wide mechanisms for addressing the widespread sources of pollution.

The success of these governmental approaches remains to be seen. Citizen action, on the other hand, lends itself particularly well to the problems caused by land-based sources of debris. Reduction of the amount of waste that has to be landfilled, barged, or incinerated rests squarely with us as consumers. Careful purchasing, trash sorting, and recycling are all part of the solution. Citizen participation in the planning of local facilities and long-range approaches to solid waste practices will be critical if future actions are to be shaped in practical ways that consumers can live with.

As for industry's role in the future, technological advancements in the fields of degradable plastics and recycling are essential steps. Fortunately, industry has stepped up its efforts in exploring these possibilities and has made significant strides toward potential solutions. But we must be careful not to view degradable plastics as a panacea to the debris problem. It is neither desirable nor feasible to make all plastics degradable. In the first place, degradability does nothing to discourage the practice of using the ocean as a garbage can. If consumers and seafarers thought all plastic products were degradable, they would have no incentive to carry them away from our beaches or to stop dumping wastes overboard. Secondly, not all plastic products lend themselves to degradable applications. Degradability should be considered primarily for those plastic products that pose the greatest harm to marine resources.

Ultimately, the solutions to the problems of plastic debris depend on continued cooperation among industry, policy makers, and the general public. At no time in history has attention to the marine debris problem been greater. We must use this momentum to confront these problem areas and eliminate their contributions to the marine debris problem.

We all have the potential to do something to reduce the problems caused by plastics in the ocean. By reading this guide and becoming aware of the problem, you have already taken the first step. Now, become part of the solution!

APPENDICES

APPENDIX A: Federal Agencies

Following are the addresses of U.S. federal departments, agencies, offices, advisory committees, and commissions mentioned in the text. The addresses of selected regional offices of these agencies are included, and some may have other offices throughout the country. To find the office nearest you, consult your telephone directory or local library, or write to the main office listed below.

DEPARTMENT OF COMMERCE

National Oceanic and Atmospheric Administration
14th Street and Constitution Avenue NW
Washington, D.C. 20230

National Marine Fisheries Service
Universal Building
1825 Connecticut Avenue, NW
Washington, D.C. 20235

National Ocean Service
National Marine Pollution Program Office
Rockwall Building, Room 610
11400 Rockville Pike
Rockville, MD 20852

National Marine Fisheries Service Regional Offices
Northeast Region
Federal Building
14 Elm Street
Gloucester, MA 01930

Southeast Region
9450 Koger Boulevard
Duval Building
St. Petersburg, FL 33702

Northwest Region
7600 Sand Point Way, NE
Seattle, WA 98115

Southwest Region
300 South Ferry Street
Terminal Island, CA 90731-7415

Alaska Region
Federal Building, Room 453
709 West Ninth Street
Juneau, AK 99802

Fishery Management Councils
New England Fishery Management Council
Suntaug Office Park
5 Broadway, Route 1
Saugus, MA 01906

Mid-Atlantic Fishery Management
Council
Federal Building, Room 2115
300 South News Street
Dover, DE 19901

South Atlantic Fishery Management Council
1 Southpark Circle
Suite 306
Charleston, SC 29407

Gulf of Mexico Fishery Management Council
Lincoln Center, Suite 881
5401 W. Kennedy Blvd.
Tampa, FL 33609

Caribbean Fishery Management
Council
Bonco de Ponce Building
Suite 1108
Hato Rey, Puerto Rico 00918

Pacific Fishery Management
Council
Metro Center, Suite 420
200 S.W. First Avenue
Portland, OR 97201

North Pacific Fishery Management
Council
411 West 4th Avenue, Suite 2-D
P.O. Box 103136
Anchorage, AK 99510

Western Pacific Fishery Management Council
1164 Bishop Street, Room 1405
Honolulu, HI 96813

MARINE MAMMAL COMMISSION
1625 I Street, NW
Suite 300
Washington, DC 20005

ENVIRONMENTAL PROTECTION AGENCY

Office of Marine and Estuarine
Protection
WH-556M
401 M Street, SW
Washington, DC 20460

Office of Pesticides and Toxic
Substances
TS-778
401 M Street, SW
Washington, DC 20460

EPA Regional Offices
REGION 1 (CT, ME, MA, NH,
RI)
Environmental Protection Agency
JFK Federal Bldg., Room 2203
Boston, MA 02203

REGION 2 (NJ, NY, PR, VI)
Environmental Protection Agency
26 Federal Plaza
New York, NY 10278

REGION 3 (DE, DC, MD, PA,
VA)
Environmental Protection Agency
841 Chestnut Street
Philadelphia, PA 19107

REGION 4 (AL, FL, GA, MS, NC, SC)
Environmental Protection Agency
345 Courtland Street, NE
Atlanta, GA 30365

REGION 5 (Great Lakes)
Environmental Protection Agency
230 S. Dearborn
Chicago, IL 60604

REGION 6 (LA, TX)
Environmental Protection Agency
1445 Ross Avenue
Dallas, TX 75202

REGION 9 (CA, HI, AMERICAN SAMOA, TRUST TERRITORIES PACIFIC GUAM, N. MARIANAS)
Environmental Protection Agency
215 Fremont Street
San Francisco, CA 94105

REGION 10 (OR, WA, AK)
Environmental Protection Agency
1200 Sixth Avenue
Seattle, WA 98101

DEPARTMENT OF AGRICULTURE

Animal and Plant Health Inspection Service
Plant Protection and Quarantine
6505 Bellcrest Road
Federal Center Building
Hyattsville, MD 20782

DEPARTMENT OF INTERIOR

Minerals Management Service
18th and C St. NW
Room 4230
Wahsington, DC 20240

DEPARTMENT OF THE NAVY

Shipboard Pollution Abatement Program OP-45
Chief of Naval Operations
Washington Navy Yard
Bldg. 200, 1st Floor, Wing 3-S
Washington, DC 20350

DEPARTMENT OF STATE

Office of Oceans and Polar Affairs
OES/OPA
Room 5801
2201 C Street, NW
Washington, DC 20520

DEPARTMENT OF TRANSPORTATION

United States Coast Guard
Boating, Public and Consumer Affairs
2100 Second Street, S.W.
Washington, DC 20250

APPENDIX B: Activities of the Federal Marine Entanglement Research Program

For information on any of the following activities contact:

Marine Entanglement Research Program
Northwest and Alaska Fisheries Center
National Marine Fisheries Service
National Oceanic and Atmospheric Administration
7600 Sand Point Way, NE
Seattle, Washington 98115-0070

For copies of educational materials or documents contact:

Marine Debris Information Office
Atlantic Coast and Gulf Coast
1725 DeSales St. NW
Washington, DC 20036

Marine Debris Information Office
Pacific Coast
312 Sutter Street, Suite 606
San Francisco, CA 94108

1985 ACTIVITIES

EDUCATION
Education Program Development and Implementation for the North Pacific Region
West Coast/New England Coast Beach Cleanup
Hawaiian Workshop on the Fate and Impact of Marine Debris

RESEARCH
Northern Fur Seal Entanglement Research
Northern Sea Lion Entanglement Research
Establishment of a Reference Collection to Identify Marine Debris
Beach Accumulation and Loss Rate Estimation in Alaska
Analysis of U.S. Fishery Observer Data on Marine Debris in the Foreign and Joint Venture Groundfish Fishery
Survey of High Seas Squid Gillnet Fishery
Identification of Sources of Fishing Debris Affecting Endangered Marine Animals in the Northwestern Hawaiian Islands
The Fate and Impact of Derelict Gillnets
Impact of Ingested Debris on Sea Turtles
Impact of Ingested Debris on Seabirds
Method of Surveying Distribution and Abundance of Debris at Sea
Expansion of Information Collected by Stranding Programs

Evaluation of Aerial Techniques for Assessing Debris Density

MITIGATION
Development of Methods to Reduce the Disposal of Ship Generated Refuse
in the Marine Environment
Research on the Use of Degradable Materials

1986 ACTIVITIES
EDUCATION
Marine Debris Education Continued and Expanded to Atlantic and Gulf of
Mexico

RESEARCH
Alaskan Beach Debris Survey Methodology
Survey of High Seas Squid Driftnet Fisheries
Hawaiian Island Endangered Species Monitoring (cont'd.)
Fur Seal Responses to Derelict Fishing Gear
Entanglement Rates of Female Northern Fur Seals
Northern Fur Seal and Sea Lion Pup Entanglement Assessment
Debris Ingestion by Hawaiian Seabirds
Dynamics of Gillnet Gear (cont'd.)
Impact of Debris on the Ocean Bottom
Ingestion of Debris by Cetaceans
Infrared Spectrophotometric Analysis of Derelict Fishing Gear

MITIGATION
Disposal Methods Development
Fur Seal Rookery Cleanup
Photodegradation Processes in the Marine Environment

1987 ACTIVITIES
EDUCATION
Marine Debris Education (cont'd)
North Pacific Education Program Evaluation
Marine Debris Teaching Unit Development for Project Wild
Development of Manual on Procedures for Monitoring Plastic Debris on
Beaches and at Sea

RESEARCH
High Seas Squid Fishery Impacts

91

Hawaiian Monk Seal Entanglement Protection and Evaluation
Dynamics of Gillnet Gear
Northern Fur Seal Entanglement Studies
Channel Islands Pinniped Entanglement Monitoring
Alaska Beach Litter Index
Sampling Survey of Impacts of Marine and Coastal Debris and
    Entanglement on Sea Turtles
Analyses of Sea Turtle Stomachs Collected from Strandings on the
    Atlantic Coast
Composition and Weathering of Derelict Trawl Web Collected from
    Alaskan Beaches
Marine Debris in Upwelling and Frontal Zones in the Gulf of Mexico
Assessement of Floating Plastic Particles
Completion of Hawaiian Seabird Plastic Ingestion Impacts
Support for Pacific Rim Fishermen's Conference on Marine Debris

MITIGATION
Plastics Research Steering Group Meeting
Assessment of Vessel Refuse Reception Problems in Alaskan Ports
Evaluation of Plastic Recycling Systems

1988 ACTIVITIES

EDUCATION
North Pacific Education Program (cont'd)
Gulf of Mexico Education Program (cont'd)
NW Atlantic Education Program (cont'd)
Second International Conference on Marine Debris

MITIGATION
Degradable Plastic Technology Research and Development
Unalaska Port Waste Management Planning
Waste Burning aboard ships: A Study of Modest Technology and Low Cost
    Systems, their Efficiency and Safety

RESEARCH
High Seas Squid Fishery Impacts
Surveys of Juvenile Male Fur Seals
Behavior of Entangled and Unentangled Juvenile Male Fur Seals
Monk Seal Protection
Channel Islands Pinniped Entanglement Monitoring
Alaska Beach Litter: Disappearance and Accumulation

1989 ACTIVITIES

EDUCATION AND PUBLIC AWARENESS
Marine Debris Information Offices (cont'd)
Marine Debris Education Program Supplies
Marine Debris Education Program for Shipping and Cruise Lines; Phase II
Second International Conference on Marine Debris (cont'd)

MITIGATION
National Marine Debris Database Support
Debris Removal from Hawaiian Monk Seal Beaches (cont'd)
Advisory Workshop on Marine Debris Mitigation
Ghost Gillnet Mitigation Experiments

IMPACTS RESEARCH AND MONITORING
High Seas Driftnet Fisheries Investigations (cont'd)
Impact of Plastic Particle Ingestion on Survival of Larval Fish (cont'd)
Effects of Marine Pollution on Juvenile, Pelagic Sea Turtles (cont'd)
Assessment of Marine Debris, Entanglement and Causes of Death in Sea
    Turtles (cont'd)
Juvenile Male Fur Seal Survivorship and Entanglement in Marine Debris
    (cont'd)
The Effects of Ingested Plastic on Albatross (cont'd)
Ingestion of Marine Plastics by Western Atlantic Seabirds
Bioaccumulation and Buoyancy of Floating Marine Debris
Report of a Marine Debris Survey of the Eastern Bering Sea Shelf
National Seashore Marine Debris Survey Program (cont'd)
Surveys of Entangling Debris on Alaska Beaches (cont'd)
Channel Islands Pinniped Entanglement Monitoring (cont'd)

REPORTS

The following is a partial listing of reports that are available from the
Marine Entanglement Research Program and the Marine Debris Information
Office:

Andrady, A.L. 1987. Research on the Use of Degradable Fishing Gear and
    Packaging Materials. NWAFC Processed Report 87-03.

Berger, J. and C. Armistead. 1986. Discarded Net Material in Alaskan Waters,
    1982–84. NOAA Technical Memorandum NMFS F/NWC-110.

Coe, J.M. and A.R. Bunn. 1987. Description and Status of Tasks in the National Oceanic and Atmospheric Administration's Marine Entanglement Research Program for Fiscal Years 1985–1987. NWAFC Processed Report 87-15.

Day, R., D. Clausen, and S. Ignell. 1986. Distribution and Density of Plastic Particulates in the North Pacific Ocean in 1986. Document submitted to the International North Pacific Fisheries Commission, Anchorage, Alaska, November 1986. 17p. (Northwest and Alaska Fisheries Center, National Marine Fisheries Service, NOAA, Auke Bay Laboratory, P.O. Box 210155, Auke Bay, AK, 99821.)

Gerrodette, T., B. Choy and L.M. Hiruki. 1987. An Experimental Study of Derelict Gill Nets in the Central Pacific Ocean. Southwest Fisheries Center Administrative Report H-87-18.

Henderson, J.R., S.L. Austin and M.B. Pillos. 1987. Summary of Webbing and Net Fragments Found on Northwestern Hawaiian Islands beaches, 1982–86. Southwest Fisheries Center Administrative Report H-87-11.

Ignell, S. and M. Dahlberg. 1986. Results of 1986 Cooperative Research on the Distribution of Marine Debris in the North Pacific Ocean. Document submitted to the International North Pacific Fisheries Commission, Anchorage, Alaska, November 1986. 16p. (Northwest and Alaska Fisheries Center, National Marine Fisheries Service, NOAA, Auke Bay Laboratory, P.O. Box 210155, Auke Bay, AK, 99821.)

Ignell, S., J. Bailey and J. Joyce. 1986 Observations on High-Seas Squid Gill-Net Fisheries, North Pacific Ocean, 1985. NOAA Technical Memorandum NMFS F/NWC-105.

Johnson, S.W. and R.R. Merrell. 1988. Entanglement Debris on Alaskan Beaches, 1986. NOAA Technical Memorandum NMFS F/NWC.

Loughlin T., P. Gearin, R. DeLong and R. Merrrick. 1986. Assessment of Net Entanglement on Northern Sea Lions in the Aleutian Islands, 25 June–15 July 1985. NWAFC Processed Report 86-02.

Merrell, T. and S. Johnson. 1986. Surveys of Plastic Litter on Alaskan Beaches, 1985. NOAA Technical Memorandum NMFS F/NWC-116.

Neilson, J. 1986. Final Report—Get the Drift and Bag It. NWAFC Processed Report 86-11.

Parker, N.R., S.C. Hunter and R.J. Yang. 1987. Development of Methodology to Reduce the Disposal of Non-degradable Refuse into the Marine Environment. Contract 85-ABC-00203.

Recht, F. 1988. Guidelines for Port-Based Projects to Mitigate Marine Debris.

Recht, F. 1988. Marine Refuse Disposal Project Report. NOAA Technical Memorandum NMFS F/NWC.

Ribic, C.A. and L.J. Bledsoe. 1986. Design of Surveys for the Density of Surface Marine Debris in the North Pacific. NWAFC Processed Report 86-12, 69p.

Videos available for loan or purchase:

"Marine Refuse Disposal Project of Newport, Oregon" 9:00 minute. Westcom Productions. NOAA/NMFS.

"Trashing the Oceans" 7:30 minute. Saltwater Productions. NOAA/NMFS.

OTHER INFORMATION

- For more information on the education plan designed for the commercial shipping industry contact:

    Kearney/Centaur Division
    A.T. Kearney, Inc.
    225 Reinekers Lane
    P.O. Box 1438
    Alexandria, VA 22313

- Copies of a handbook on procedures for determining whether strandings of marine mammals were natural or from human-induced causes such as ingestion and entanglement in debris, are available from the authors at the Marine Mammal Program at the National Museum of Natural History, Smithsonian Institution, Washington, DC 20560 referenced as:

    Hare, M.P. and J.G. Mead. 1987. Handbook for Determination of Adverse Human-Marine Mammal Interactions from Necropsies. NWAFC Processed Report 87-06, 35p.

- A reference collection to help identify marine debris has been established at the Northwest and Alaska Fisheries Center (NWAFC) in Seattle, Washington. Thus far techniques have been developed to identify various types of plastics used to manufacture fishing gear. The staff at NWAFC is accepting materials for identification and may be contacted for information on procedures for delivery of specimens.

APPENDIX C: Sea Grant Offices

The National Sea Grant Program was established to take advantage of the knowledge and research at the nation's universities in solving marine problems and using marine resources wisely. Each Sea Grant Program across the country is based at a university and committed to serve national and local constituencies in the marine community through research, education, and advisory services. Many Sea Grant programs have begun to develop activities and materials directed at marine debris. A listing of Sea Grant Programs follows:

Alabama Sea Grant Marine Advisory Program
3940 Government Boulevard, Suite 5
Mobile, AL 36609

Alaska Sea Grant College Program
University of Alaska
138 Irving III
Fairbanks, AK 99775-5040

Alaska Sea Grant Marine Advisory Program
University of Alaska
P.O. Box 103160
Anchorage, AK 99510

California Sea Grant Program, A-032
University of California, San Diego
La Jolla, CA 92093

University of Southern California Sea Grant and Marine Advisory
    Programs
Institute for Marine and Coastal Studies
University of Southern California
University Park
Los Angeles, CA 90089-0341

California Sea Grant
Cooperative Extension Service
University of California
Davis, CA 95616

Connecticut Sea Grant Program
Marine Sciences Institute
University of Connecticut
Avery Point
Groton, CT 06340

Delaware Sea Grant Program
University of Delaware
Robinson Hall
Newark, DE 19716

Delaware Sea Grant Program
College of Marine Studies
University of Delaware
700 Pilottown Road
Lewes, DE 19958

Florida Sea Grant College Program
University of Florida
Building 803
Gainesville, FL 32611

Florida Sea Grant Extension Program
University of Florida
117 Newins/Ziegler Hall
Gainesville, FL 32611

Georgia Sea Grant College Program
University of Georgia
Ecology Building
Athens, GA 30602

Georgia Sea Grant Program
Marine Extension Service
University of Georgia
P.O. Box Z
Brunswick, GA 31523

Hawaii Sea Grant College Program
University of Hawaii
1000 Pope Road, Room 220
Honolulu, HI 96822

Illinois/Indiana Sea Grant Program
Purdue University
Department of Forestry and Natural Resources
West Lafayette, IN 47907

Illinois/Indiana Sea Grant Extension Program
University of Illinois
1010 Jorie Boulevard, Suite 300
Oak Brook, IL 60521

Louisiana Sea Grant College Program
Center for Wetlands Resources
Louisiana State University
Baton Rouge, LA 70803-7507

University of Maine/University of New Hampshire Sea Grant Program
University of Maine
14 Coburn Hall
Orono, ME 04469

Maryland Sea Grant College Program
University of Maryland
H.J. Patterson Hall, Room 1222
College Park, MD 20742

MIT Sea Grant College Program
Massachusetts Institute of Technology
77 Massachusetts Avenue
Building E38, Room 330
Cambridge, MA 02139

Sea Grant Program
Woods Hole Oceanographic Institution
Woods Hole, MA 02543

Michigan Sea Grant College Program
University of Michigan
1st Building, Room 4103
2200 Bonisteel Boulevard
Ann Arbor, MI 48109-2099

Michigan Sea Grant Program
Marine Advisory Service
Michigan State University
334 Natural Resources Building
East Lansing, MI 48824

Minnesota Sea Grant Program
University of Minnesota
116 Classroom Office Building
1994 Buford Avenue
St. Paul, MN 55108

Minnesota Sea Grant Extension Program
University of Minnesota
208 Washburn Hall
Duluth, MN 55812

Mississippi-Alabama Sea Grant Consortium
P.O. Box 7000
Ocean Springs, MS 39564-7000

Mississippi Sea Grant Marine Advisory Program
4646 W. Beach Boulevard, Suite 1-E
Biloxi, MS 39531

University of New Hampshire/University of Maine Sea Grant College Program
University of New Hampshire
Marine Program Building
Durham, NH 03824

New Jersey Marine Sciences Consortium
Sandy Hook Field Station
Building 22
Fort Hancock, NJ 07732

New Jersey Sea Grant Extension Service
Rutgers University/Cook College
P.O. Box 231
New Brunswick, NJ 08903

New York Sea Grant Institute
State University of New York at Stony Brook
Stony Brook, NY 11794-5000

New York Sea Grant Extension Program
Cornell University
12 Fenow Hall
Ithaca, NY 14853-3001

North Carolina Sea Grant Program
North Carolina State University
Box 8605
Raleigh, NC 27695-8605

Ohio Sea Grant Program
1314 Kinnear Road
Columbus, OH 43212

Oregon Sea Grant College Program
Oregon State University
Administrative Services Building-A320
Corvallis, OR 97331

Puerto Rico Sea Grant Program
University of Puerto Rico
Department of Marine Science
RUM UPR, P.O. Box 5000
Mayaguez, PR 00709-5000

Rhode Island Sea Grant College Program
University of Rhode Island
Narragansett Bay Campus
Marine Resources Building
Narragansett, RI 02882

South Carolina Sea Grant Consortium
287 Meeting Street
Charleston, SC 29401

Texas Sea Grant College Program
Texas A&M University
College Station, TX 77843

Virginia Sea Grant College Program
University of Virginia
Madison House - 170 Rugby Road
Charlottesville, VA 22903

Virginia Sea Grant Marine Advisory Program
Virginia Institute of Marine Science
Gloucester Point, VA 23062

Washington Sea Grant College Program
University of Washington, HG-30
3716 Brooklyn Avenue, NE
Seattle, WA 98105

Wisconsin Sea Grant Institute
University of Wisconsin-Madison
1800 University Avenue
Madison, WI 53705

APPENDIX D: State Agencies

The following agencies should be contacted for information on marine debris and programs directed at solving this problem in your state. If an agency does not have a current program, encourage them to become involved in solving the problem.

Alabama Department of Conservation and Natural Resources
64 North Union Street
Montgomery, AL 36130

Alabama Department of Conservation and Natural Resources
Division of Marine Resources
P.O. Box 189
Dauphin Island, AL 36528

Alabama Department of Environmental Management
1751 Federal Drive
Montgomery, AL 36130

Alaska Department of Public Safety
Fish and Wildlife Protection
5700 Tudor Road
Anchorage, AK 99507

Alaska Department of Fish and Game
Habitat Protection Division
P.O. Box 3-2000
Juneau, AK 99802

Alaska Department of Environmental Conservation
Litter Coordinator
Division of Environmental Quality
Juneau, AK 99811

California Department of Fish and Game
1416 Ninth Street
Sacramento, CA 95814

California Coastal Commission
631 Howard Street
San Francisco, CA 94105

Connecticut Department of Environmental Protection
Coastal Areas Management
71 Capitol Avenue
Hartford, CT 06106

Delaware Department of Natural Resources
Office of Information and Education
89 Kings Highway
P.O. Box 1401
Dover, DE 19903

Florida Department of Natural Resources
Division of Marine Resources
Marjory Stoneman Douglas Building
Tallahassee, FL 32303

Georgia Department of Natural Resources
Floyd Towers East
205 Butler Street
Atlanta, GA 30334

Environmental Protection Agency
P.O. Box 2999
Agana, Guam 96910

Hawaii Department of Land and Natural Resources
Box 621
Honolulu, HI 96809

Illinois Environmental Protection Agency
2200 Churchill Road
Springfield, IL 62706

Illinois Department of Conservation
Lincoln Tower Plaza
524 S. Second Street
Springfield, IL 62706

Indiana Department of Natural Resources
608 State Office Building
Indianapolis, IN 46204

Louisiana Department of Natural Resources
P.O. Box 44487
Baton Rouge, LA 70804

Maine Department of Marine Resources
State House
Station #21
Augusta, ME 04333

Maryland Department of Natural Resources
Tidewater Administration
Tawes State Office Building
Annapolis, MD 21401

Massachusetts Department of Fisheries, Wildlife and Environmental Law
   Enforcement
100 Cambridge Street
Boston, MA 02202

Massachusetts Department of Environmental Management
100 Cambridge Street
Boston, MA 02202

Michigan Department of Natural Resources
Box 30028
Lansing, MI 48909

Minnesota Department of Natural Resources
500 Lafayette Road
St. Paul, MN 55155

Mississippi Department of Wildlife Conservation
Bureau of Marine Resources
Southport Mall
P.O. Box 451
Jackson, MS 39205

New Hampshire Fish and Game Department
34 Bridge Street
Concord, NH 03301

New Jersey Department of Environmental Protection
Division of Coastal Resources
CN 401
Trenton, NJ 08625

New York Department of Environmental Conservation
50 Wolf Road
Albany, NY 12233

New York Department of Environmental Conservation
Hunters Point Plaza
Long Island City, NY 11101

North Carolina Department of Natural Resources and Community
  Development
P.O. Box 27687
Raleigh, NC 27611

Ohio Department of Natural Resources
Office of Litter Control
Fountain Square
Columbus, OH 43224

Oregon Department of Fish and Wildlife
P.O. Box 59
Portland, OR 97207

Pennsylvania Department of Environmental Resources
Press Office, 9th Floor
Fulton Building
Box 2063
Harrisburg, PA 17120

Puerto Rico Department of Natural Resources
P.O. Box 5887
Puerta de Tierra Station
San Juan, PR 00906

Rhode Island Department of Environmental Management
9 Hayes Street
Providence, RI 02908

South Carolina Wildlife and Marine Resources Department
Rembert C. Dennis Building
P.O. Box 167
Columbia, SC 29202

Texas General Land Office
Stephen F. Austin State Office Building
1700 N. Congress Ave., Room 620
Austin, TX 78701

Texas Parks and Wildlife Department
4200 Smith School Road
Austin, TX 78744

Virginia Department of Conservation and Historic Resources
1100 Washington Building
Capitol Square
Richmond, VA 23219

Virginia Marine Resources Commission
P.O. Box 756
2401 West Avenue
Newport News, VA 23607

Department of Conservation and Cultural Affairs
P.O. Box 4399
St. Thomas, VI 00801

Washington Department of Natural Resources
Public Lands Building
Olympia, WA 98504

Wisconsin Department of Natural Resources
Box 7921
Madison, WI 53707

APPENDIX E:
COASTWEEKS Participants

COASTWEEKS is a citizen's network of organizations, agencies and individuals that foster public awareness of the great value of the nation's coasts and shores during a period in early fall that includes the Columbus Day holiday. In 1989, COASTWEEKS will be extended over a three-week period from September 16–October 9. Beach cleanups have become an annual event in several states, where citizens have the opportunity to come to the beach for a day to collect marine debris. Working together, the Coastal States Organization and the Center for Marine Conservation hope to obtain a congressional resolution that declares the third Saturday in September, starting with September 23, 1989, National Beach Cleanup Day.

National steering committee members for COASTWEEKS are the Center for Marine Conservation, Coast Alliance, Coastal States Organization, League of Women Voters, National Association of State Universities and Land-Grant Colleges, Sierra Club Coastal Committee, and The Coastal Society. Following is a list of state agencies and organizations participating in COASTWEEKS 1989 (Source: Coastal States Organization):

Alabama Department of Economic and Community Affairs
Box 2939
Montgomery, AL 36105-2399

Alaska Coastal Management Program
Division of Government Coordination
Pouch AW
Juneau, AK 99811-0615

Coastal Management Program
Development Planning Office
Pago Pago, American Samoa 96799

California Coastal Commission
631 Howard Street
San Francisco, CA 94105

California Sea Grant Program, A-032
University of California-San Diego
La Jolla, CA 92093

Center for Marine Conservation
312 Sutter Street, Suite 316
San Francisco, CA 94108

University of Southern California Sea Grant Program
University Park
Los Angeles, CA 90089-1231

Commonwealth of the Northern Marianas Islands
Coastal Resources Management Office
Nauru Bldg. 6th Floor
Saipan, CNMI 96950

Connecticut Department of Environmental Protection
Coastal Areas Management
18-20 Trinity St.
Hartford, CT 06106

Connecticut Sea Grant Program
Marine Sciences Institute
University of Connecticut
Groton, CT 06340

Delaware Department of Natural Resources
Office of Information and Education
89 Kings Highway
P.O. Box 1401
Dover, DE 19903

University of Delaware College of Marine Studies and Delaware Sea Grant
    College Program
University of Delaware
263 E. Main Street
Newark, DE 19716

Center for Marine Conservation
One Beach Dr. SE
St. Petersburg, FL 33701

Florida Department of Environmental Regulation
2600 Blair Stone Road
Tallahassee, FL 32399

Florida Sea Grant Program
University of Florida
G022 McCarty Hall
Gainesville, FL 32611

Georgia Sea Grant College Program
University of Georgia
Ecology Building
Athens, GA 30602

Georgia Marine Extension Service
P.O. Box 13687
Savannah, GA 31416

Bureau of Planning
Coastal Management Program
P.O. Box 2950
Agana, Guam 96910

Hawaii Sea Grant Program
University of Hawaii
1000 Pope Road, Room 220
Honolulu, HI 96822

Louisiana Sea Grant Program
Center for Wetlands Resources
Louisiana State University
Baton Rouge, LA 70803-7507

Louisiana Department of Natural Resources
P.O. Box 44487
Baton Rouge, LA 70804-4487

Maine State Planning Office
Station 38
Augusta, ME 04333

Maine Marine Advisory Program
University of Maine
30 Coburn Hall
Orono, ME 04469

Maryland Department of Natural Resources
Coastal Resources Advisory Committee
Tawes State Office Building C-3
Annapolis, MD 21401

Maryland Sea Grant Program
University of Maryland
H.J. Patterson Hall, Room 1222
College Park, MD 20742

Massachusetts Coastal Zone Management Program
100 Cambridge Street, 20th Floor
Boston, MA 02202

MIT Sea Grant Program
Massachusetts Institute of Technology
Building E38, Room 374
Cambridge, MA 02142

National Marine Fisheries Service
Habitat Protection Branch
14 Elm Street
Gloucester, MA 09130

Michigan Department of Natural Resources
Division of Land and Water Management
Box 30028
Lansing, MI 48909

Michigan Sea Grant Program
University of Michigan
1st Building, Room 4103
2200 Bonisteel Blvd.
Ann Arbor, MI 48109-2099

Mississippi Department of Wildlife Conservation BMR
Drawer 959
Long Beach, MS 39560

Mississippi-Alabama Sea Grant Consortium
Box 7000
Ocean Springs, MS 39564

New Hampshire Office of State Planning
Coastal Program
152 Court Street
Portsmouth, NH 03801

New Jersey Department of Environmental Protection
Coastal Program
CN 401
Trenton, NJ 08625

New Jersey Sea Grant Program
New Jersey Marine Sciences Consortium
Sandy Hook Field Station
Building 22
Fort Hancock, NJ 07732

New York Department of State
Coastal Management Program
162 Washington Ave
Albany, NY 12031

New York Sea Grant Institute
State University of New York at Stony Brook
Stony Brook, NY 11794-5001

North Carolina Department of Natural Resources and Community
    Development
Division of Coastal Management
Box 27687
Raleigh, NC 27611

North Carolina Sea Grant Program
North Carolina State University
Box 8605
Raleigh, NC 27695-8605

Ohio Department of Natural Resources
Division of Water
Fountain Square
Columbus, OH 43224

Ohio Sea Grant Program
Ohio State University
1314 Kinnear Road
Columbus, OH 43212

Oregon Department of Land Conservation and Development
Coastal Program
1175 Court Street NE
Salem, OR 97223

Oregon Department of Fish and Wildlife
P.O. Box 59
Portland, OR 97207

Oregon Sea Grant Program
Oregon State University
Corvallis, OR 97331

Pennsylvania Department of Environmental Resources
Division of Coastal Zone Management
Box 1467
Harrisburg, PA 17120

Puerto Rico Department of Natural Resources
Box 5887
Puerta de Tierra
San Juan, PR 00906

Puerto Rico Sea Grant Program
University of Puerto Rico
Department of Marine Science
Box 5000
Mayaguez, PR 00709-5000

Rhode Island Sea Grant Program
University of Rhode Island
Narragansett Bay Campus
Marine Resources Building
Narragansett, RI 02882

South Carolina Coastal Council
4280 Executive Place North
Charleston, SC 29405

South Carolina Sea Grant Consortium
287 Meeting Street
Charleston, SC 29401

Center for Marine Conservation
Gulf Coast States Regional Office
1201 West 24th Street
Austin, TX 78705

Texas General Land Office
Stephen F. Austin Building, Room 620
1700 North Congress Avenue
Austin, TX 78701

Texas Sea Grant Program
Texas A&M University
College Station, TX 77843

Center for Marine Conservation
12 Cantamar Court
Hampton, VA 23664

Council on the Environment
903 9th Street
Richmond, VA 23219

Virginia Sea Grant Marine Advisory Program
Virginia Institute of Marine Science
Gloucester Point, VA 23062

Virgin Islands Marine Advisory Service
University of the Virgin Islands
St. Thomas, VI 00802

Department of Planning and Natural Resources
#179 Altona and Welgunst
St. Thomas, VI 00801

Washington Department of Ecology-Shorelands Division
Mail Stop PV-11
Olympia, WA 98504

Washington Sea Grant Program
University of Washington, HG-30
3716 Brooklyn Avenue, NE
Seattle, WA 98105

Wisconsin Coastal Management Program
101 S. Webster Street
6th Floor
Madison, WI 53707

Wisconsin Sea Grant Institute
University of Wisconsin
1800 University Avenue
Madison, WI 53705

APPENDIX F: 1988 Beach Cleanups Coordinators by State

During COASTWEEKS and throughout the year every coastal state conducts coastal cleanup activities. The following persons volunteer their time as state coordinators, helping volunteers find their local cleanup and distributing materials to area "zone captains." In many cases these persons have become the local experts on beach cleanups and statewide cleanup activities.

Alabama
John Marshall
Alabama Department of Environmental Management
2204 Perimeter Road
Mobile, AL 36615
(205) 479-2336

Alaska
Audrey Lee
Alaskans For Litter Prevention and Recycling
P.O. Box 231231
Anchorage, AK 99523
(907) 272-9326

California
Jack Liebster
California Coastal Commission
631 Howard Street
San Francisco, CA 94105
(415) 543-8555

Connecticut
Peg Van Patten
Connecticut Sea Grant Program
University of Connecticut
Marine Science Institute
Avery Point
Groton, CT 06340
(203) 445-3459

Delaware
Donna Stachecki Sharpe
Information Officer
Delaware Department of Natural Resources and Environmental Control
89 Kings Highway
P.O. Box 1401
Dover, DE 19901
(302) 736-4506

Florida
Edward Proffitt
Center for Marine Conservation
Bayfront Tower
1 Beach Drive SE, Suite 304
St. Petersburg, FL 33701
(813) 895-2188

Georgia
Jay Calkins
Marine Extension Service
P.O. Box 13687
McWhorder Drive
University of Georgia, Skidaway Island
Savannah, GA 31416
(912) 356-2496

Hawaii
John Yamauchi
Hawaii State Litter Control Office
205 Koula Street
Honolulu, HI 96813
(808) 548-3400 or 548-6444

Louisiana
Barbara Colthrap
Louisiana Department of Culture, Recreation and Tourism
P.O. Box 94291
Baton Rouge, LA 70804-9291
(504) 342-8148

Jim Hanifen
Baton Rouge Audubon Society
3343 Myrtle Avenue
Baton Rouge, LA 70806
(504) 765-2390

Maine
Flis Schauffler
Maine Coastal Program
State Planning Office
184 State Street
State House Station 38
Augusta, ME 04333
(207) 289-3261

Maryland (see Virginia)

Massachusetts
Anne Smrcina
Jane Alford
Coastal Zone Management Program
100 Cambridge Street, 20th Floor
Boston, MA 02202
(617) 727-9530

Thomas Bigford
National Marine Fisheries Service
Habitat Conservation Branch
2 State Fish Pier
Gloucester, MA 01930-3097
(617) 281-3600 ex. 209

Mississippi
Jim Franks and Dianne Hunt
Mississippi Bureau of Marine Resources
P.O. Drawer 959
Long Beach, MS 39560
(601) 864-4602

Sharon Walker
MS-AL Sea Grant Consortium
P.O. Box 7000
Ocean Springs, MS 39564
(601) 875-9341

Gail Bishop
Gulf Islands National Seashore
3500 Park Road
Ocean Springs, MS 39564
(601) 875-0074

New Hampshire
Julia Steed Mawson
Visitor's Center at Odiorne Point
c/o UNH Sea Grant
MEC Administration Building
University of New Hampshire
Durham, NH 03824
(603) 436-8043 or (603) 862-3460

New Jersey
Department of Environmental Protection
Division of Coastal Resources
501 East State Street
CN 401
Trenton, NJ 08625
(609) 292-8973

Patricia Morton-Toth
Alliance for a Living Ocean
P.O. Box 95
Ship Bottom, NJ 08008
(609) 698-7966

Valarie Maxwell
Clean Ocean Action
P.O. Box 505
Building 18 Hartshorne Drive
Sandy Hook
Highlands, NJ 07732
(201) 872-0111

New York
Roberta Weisbrod
New York State Department of Environmental Conservation
Hunters Point Plaza
Long Island City, NY 11101
(718) 482-4992

North Carolina
Northern Beaches: Rich Novak
UNC Sea Grant
North Carolina Aquarium/Roanoke Island
Airport Road
Manteo, NC 27954
(919) 473-3937

Middle Beaches: Diane Warrender
Keep America Beautiful
Court House Square
Beaufort, NC 28516
(919) 728-8595

Southern Beaches: Andy Wood
North Carolina Aquarium/Fort Fisher
Kure Beach, NC 28449
(919) 458-8257

Oregon
Judie Neilson
Oregon Department of Fish and Game
P.O. Box 59
Portland, OR 97207
(503) 229-5406

Pennsylvania
Genny Volgstadt
Department of Environmental Resources
Preske Island State Park
P.O. Box 8510
Erie, PA
(814) 871-4251

Puerto Rico
Ruperto Chaparro
UPR Sea Grant Program
RUM-UPR
P.O. Box 5000
Mayaguez, PR 00709-5000
(809) 832-8045

Rhode Island
Eugenia Marks
Audubon Society of Rhode Island
12 Sanderson Road
Smithfield, RI 02917
(401) 231-6444

Michelle Merola
Department of Environmental Management
OSCAR Program
83 Park Street
Providence, RI 02903
(401) 277-3434

South Carolina
Virginia Beach
South Carolina Sea Grant Consortium
287 Meeting Street
Charleston, SC 29401
(803) 727-2078

Texas
Linda Maraniss
Center for Marine Conservation
1201 West 24th Street
Austin, TX 78705
(512) 477-6424

Angela Farias
Texas General Land Office
1700 N. Congress Avenue
Austin, TX 78701
1-800-85BEACH or (512) 463-5108

Virginia
Ocean Beaches
Edward Risley
Audubon Naturalist Society
7212 Beachwood Road
Alexandria, VA 22307
(703) 768-8478

David Cottingham
U.S. Department of Commerce
National Oceanic and Atmospheric Adminstration
Office of Chief Scientist
14th Street and Constitution Avenue
Room 6222
Washington, DC 20230
(202) 377-5181

Neil FitzPatrick
Audubon Naturalist Society
8940 Jones Mill Road
Chevy Chase, MD 20815
(301) 652-5964

Fran Krieg
Assateague Mobile Sportsfishermen's Association
P.O. Box 149
Ocean City, MD 21842
(301) 957-9971

Judy Johnson
Committee to Preserve Assateague Island, Inc.
616 Piccadilly Road
Towson, MD 21204
(301) 828-4520

Chesapeake Bay (Hampton)
Candy Tomlinson
Sierra Club
216 Susan Constant Drive
Newport News, VA 23602
(804) 764-7633

Susan Larson
Hampton Clean City Commission
22 Lincoln Street
Hampton, VA 23669
(804) 727-6394

Virgin Islands
Natalie Peters
Virgin Islands Marine Advisory
University of the Virgin Islands
St. Thomas, VI 00802
(809) 776-9200 x-1242

Patricia Mortenson
Environmental Studies Teacher
Government of the Virgin Islands
Department of Education
Charlotte Amalie
St. Thomas, VI 00801

Washington
Betsy Peabody and Ken Pritchard
Puget Sound Bank
Washington Adopt-A-Beach
607 3rd Avenue, Room 210
Seattle, WA 98104
(206) 296-6544

Pacific Northwest 4-Wheel Drive Association's Operation Shore Patrol
Camille Johnson
Washington State Parks and Recreation Commission
7150 Cleanwater Lane
Olympia, WA 98504
(206) 753-5759

APPENDIX G: Stranding Network Participants

Since all species of marine mammals are protected under the Marine Mammal Protection Act, only authorized persons may handle stranded animals, including those that have become entangled in marine debris or incidentally captured in fishing gear. The National Marine Fisheries Service has granted permits to a number of individuals and organizations that respond to such incidents concerning marine mammals. Similarly, sea turtles are protected under the Endangered Species Act and only authorized persons may handle stranded turtles under permit from the U.S. Fish and Wildlife Service. These federally permitted agencies or organizations in turn act as coordinators of groups within their regions who respond to stranded animals. Collectively these groups are referred to as "Stranding Networks" (not to be confused with the Entanglement Network). Stranding Networks are becoming an important source of information on entanglement of marine wildlife in debris or the incidental capture of animals in fishing gear. Contact these persons to report marine mammal or sea turtle strandings or for more information on their activities. (Note: * = responds to marine mammal strandings, + = responds to sea turtle strandings.)

Alabama
Robert Shipp, Ph.D. +
University of Alabama
Department of Biology
Mobile, AL 36688
(205)460-6351

Alaska
Steve Zimmerman or
John Sease*
Office of Marine Mammals and Endangered Species
Nat'l Marine Fisheries Service
Alaska Region
Juneau, AK 99802
(907)586-7233

California

Elizabeth Jozwiak*
Nat'l Marine Fisheries Service
300 S. Ferry Street
Terminal Island, CA 90731
(213)514-6199

or the local office of the CA Dept. of Fish and Game

Connecticut

Neil Overstrom* +
Mystic Marinelife Aquarium
Mystic, CT 06355
(203)536-9631

Delaware

Delaware Marine Police* +
Div. of Fish and Wildlife
P.O. Box 1401
Dover, DE 19903
(302)736-4580

Florida

Ellie Roche*
Permit Specialist
Nat'l Marine Fisheries Service
9450 Koger Blvd.
St. Petersburg, FL 33702
(813)893-3366

Daniel K. Odell, Ph.D.*
SEUS Scientific Coordinator and FL State Coordinator
Sea World Research Institute
Florida Marine Science Center
P.O. Box 590421
Orlando, FL 32859-0471
(305)345-5120

Walt Conley+
FL Dept. of Natural Resources
Bureau of Marine Research
100 Eighth Avenue, SE
St. Petersburg, FL 33701
(813)896-8626

Manatee Strandings:
In Florida—FL Dept. of Natural Resources: 1-800-342-1821
Outside Florida—U.S. Fish & Wildlife Service: (904)372-2571

Georgia
Arnold Woodward+
GA Dept. of Natural Resources
Marine Resources Division
1200 Glynn Avenue
Brunswick, GA 21523
(912)264-7218

Hawaii
State of Hawaii* +
Division of Conservation and Resources Enforcement
24-hour hotline:
in Oahu: (808)548-5918
outside Oahu: dial "0" and ask for "Enterprise 5469."

Other agencies that may be contacted include:

NMFS/Honolulu Laboratory* +
2570 Dole Street
Honolulu, HI 96822-2396
(808)943-1221

NMFS/Law Enforcement Branch* +
(808)541-2727

U.S. Fish and Wildlife Service Enforcement Branch* +
(808)541-2681

Louisiana
Steve Rabalais +
LA Univ. Marine Consortium
Marine Resources and Education Center
Chauvin, LA 70344
(504)851-2808

Maine
Bob Gowell* +
Nat'l Marine Fisheries Service
Law Enforcement Division
U.S. Courthouse
156 Federal Street
Portland, ME 04101
(207)780-3241

Massachusetts
Jeff Boggs* +
New England Aquarium
Central Wharf
Boston, MA 02110
(617)973-5247

George King*
Sealand
Brewster, MA 02631
(617)385-9252

Robert Prescott +
MA Audubon Society
Wellfleet Bay Wildlife Sanctuary
P.O. Box 236
South Wellfleet, MA 02663
(617)349-2615

Mississippi
Dr. Moby A. Solangi*
Marine Animal Productions
Box 4078
Gulfport, MS 39502
(601)864-2511

Ted Simon, Ph.D. +
Gulf Islands National Seashore
3500 Park Road
Ocean Springs, MS 39564
(601)875-9057

New Jersey
Robert Schoelkopf* +
Marine Mammal Stranding Center
P.O. Box 773
Brigantine, NJ 08203
(609)266-0538 or 348-5018

New York
Sam Sadove* +
Okeanos Foundation
216 E. Montauk Highway
Hampton Bays, NY 11946
(516)728-4522 or 728-4523

North Carolina
William J. Bowen*
NMFS Beaufort Laboratory
Beaufort, NC 28516-9722
(919)728-8740

Tom Henson +
NC Wildlife Resources Comm.
Route 1, Box 724B
Chocowinity, NC 27817
(919)946-1969

Oregon
Marine mammal strandings should be reported to the Oregon State Police

Puerto Rico
Kathy Hall +
University of Puerto Rico
Department of Marine Science
RUM-UPR P.O. Box 5000
Mayaguez, PR 00709-5000
(809)834-4040 ext. 2511
or (809)872-6513

Rhode Island
C. Robert Shoop, Ph.D. +
University of Rhode Island
Department of Zoology
Kingston, RI 02882
(401)792-2372

South Carolina
Dr. Albert E. Sanders*
The Charleston Museum
360 Meeting Street
Charleston, SC 29402
(803)722-2996

Sally Murphy +
SC Wildlife and Marine Resources Department
P.O. Box 12559
Charleston, SC 29412
(803)795-6350

Texas
Dr. Raymond Tarpley*
Dept. of Veterinary Anatomy
Texas A&M University
College Station, TX 77843
(409)845-4344

Robert G. Whistler +
Padre Island National Seashore
9405 South Padre Island Drive
Corpus Christi, TX 78418
(512)949-8173

Virginia
Jack Musick, Ph.D.* +
VA Inst. of Marine Science
School of Marine Sciences
Gloucester, VA 23062
(804)642-7317

Washington
Marine mammal strandings should be reported to the Washington State Patrol

APPENDIX H: Entanglement Network

If you are a member of an environmental organization you may already be part of a larger group that is addressing the plastic debris problem. Recognizing the severe threats that entanglement poses to marine species, more than twenty national and international organizations have joined forces as the **Entanglement Network** to share information and organize ranks to attack this problem.

American Cetacean Society
National Headquarters
P.O. Box 2639
San Pedro, CA 90731

American Humane Association
322 Massachusetts Ave, NE
Washington, DC 20002

Animal Protection Institute of America
P.O. Box 57006
Washington, DC 20037

California Marine Mammal Center
Marin Headlands
Golden Gate National Recreation Area
Fort Cronkhite, CA 94965

Center for Coastal Studies
Cetacean Research Program
59 Commercial Street, Box 826
Provincetown, MA 02657

Center for Marine Conservation
(formerly The Center for Environmental Education)
1725 DeSales Street, NW Suite 500
Washington, DC 20036

Cetacean Society International
P.O. Box 9145
Wethersfield, CT 06109

Defenders of Wildlife
1244 19th Street, NW
Washington, DC 20036

Environmental Defense Fund
1616 P Street, NW Suite 150
Washington, DC 20036

Friends of Animals
400 First Street, NW
Washington, DC 20001

Friends of the Sea Otter
P.O. Box 221220
Carmel, CA 93922

Greenpeace U.S.A.
1436 U Street, NW
Washington, DC 20009

HEART
Box 681231
Houston, TX 77268-1231

The Humane Society of the United States
2100 L Street, NW
Washington, DC 20037

International Wildlife Coalition
1807 H Street, NW #301
Washington, DC 20006

Maine Audubon Society
Gilsland Farm
118 US Route 1
Falmouth, ME 04105

Monitor Consortium
1506 19th Street, NW
Washington, DC 20036

National Audubon Society
801 Pennsylvania Avenue, SE
Suite 301
Washington, DC 20003

National Resources Defense Council
1350 New York Avenue, 3rd. Floor
Washington, DC 20005

National Wildlife Federation
1412 16th Street, NW
Washington, DC 20036-2266

Northwind Undersea Institute
1725 N Street, NW
Washington, DC 20036

The Oceanic Society
1536 16th Street, NW
Washington, DC 20036

Society for Animal Protection Legislation
P.O. Box 3719
Georgetown Station
Washington, DC 20007

The Whale Center
3929 Piedmont Avenue
Oakland, CA 94611

World Wildlife Fund
1250 24th Street, NW
Washington, DC 20037

APPENDIX I: Additional Organizations

The following organizations can be contacted for more information on marine debris and their role in solving this problem:

American Association of Port Authorities
1010 Duke Street
Alexandria, VA 22314

American Institute of Merchant Shipping
1625 K Street, NW Suite 1000
Washington, DC 20006

American Petroleum Institute
2101 L Street, NW
Washington, DC 20037

Center for Plastics Recycling Research
Busch Campus
Building 3529
Piscataway, NJ 08855

Coastal States Organization
444 N. Capitol Street, NW
Washington, DC 20001

Council of American Flag Ship Operators
1627 K Street, NW
Washington, DC 20006

Dow Chemical Company
Plastics Public Affairs Department
2040 Willard H. Dow Center
Midland, MI 48674

National Association for Plastics Container Recovery
5024 Parkway Plaza Boulevard
Suite 200
Charlotte, NC 28217

National Fisheries Institute
2000 M Street, NW Suite 580
Washington, DC 20036

National Ocean Industries Association
1050 17th Street, NW
Washington, DC 20036

Offshore Operators Committee
P.O. Box 50751
New Orleans, LA 70150

Plastics Recycling Foundation
1275 K Street NW
Suite 400
Washington, DC 20005

Port of Newport
600 SE Bay Boulevard
P.O. Box 1065
Newport, OR 97365

Sport Fishing Institute
1010 Massachusetts Avenue, NW Suite 100
Washington, DC 20001

Tampon Applicator Creative Klubs International (TACKI)
P.O. Box 819
Provincetown, MA 02657

The Society of the Plastics Industry
1275 K Street NW, Suite 400
Washington, DC 20005

- Council on Plastic Packaging and the Environment (COPPE)
  1275 K Street NW, Suite 400
  Washington, DC 20005

- Council for Solid Waste Solutions
  1275 K Street NW
  Suite 300
  Washington, DC 20005

- Plastic Bottle Institute
  Division of the Society of the Plastics Industry
  355 Lexington Avenue
  New York, NY 10017

APPENDIX J: International Organizations

The following organizations, which are referred to in the text, have information on plastic debris.

Secretariat
Law of the Sea Treaty
United Nations
Room 1827 A
New York, NY 10017

Regional Seas Activity Center
UNEP
Palais de Nations
1121 Geneva 10
Switzerland

International Maritime Organization
4 Albert Embankment
London SE1 7SR
England
United Kingdom

APPENDIX K: Laws and Treaties

Following are laws and treaties mentioned in the text. The full text of these documents or summaries of their provisions can be reviewed in the United States Code, U.S. Treaties in Force, or other references available at most public libraries. Copies of laws, treaties, and implementing regulations may frequently be obtained from the main office of the federal agency responsible for enforcement. Copies of these documents also may be obtained from the Government Printing Office, Superintendent of Documents, Washington, DC 20402, for a small fee.

UNITED STATES LAWS

Act to Prevent Pollution from Ships (33 U.S.C. 1901)
Clean Water Act (33 U.S.C. 1251)
Coastal Zone Management Act (16 U.S.C. 1451)
Comprehensive Environmental Response, Compensation and Liability Act (42 U.S.C. 9601)
Deepwater Port Act (33 U.S.C. 1501)
Endangered Species Act (16 U.S.C. 1531)
Fishery Conservation and Management Act (16 U.S.C. 1801)
Marine Mammal Protection Act (16 U.S.C. 1361)
Marine Plastic Pollution Research and Control Act (33 U.S.C. 1901)
Marine Protection, Research and Sanctuaries Act (33 U.S.C. 1401)
Ocean Pollution Planning Act (33 U.S.C. 1701)
Outer Continental Shelf Act (43 U.S.C. 1301)
Resource Conservation and Recovery Act (42 U.S.C. 6901)
Rivers and Harbors Act of 1899 (33 U.S.C. 407)
Toxic Substances Control Act (15 U.S.C. 2601)

INTERNATIONAL TREATIES AND CONVENTIONS

Convention for the protection and development of the marine environment of the wider Caribbean region. Done at Cartagena, 1983. 22 I.L.M. 227(1983).

Convention on the prevention of marine pollution by dumping from ships and aircraft. Done at Oslo, Feb. 15, 1972. 11 I.L.M. 262(1972).

Convention on the prevention of marine pollution by dumping of wastes and other matter. Done at London, Dec. 29, 1972. 26 U.S.T. 2406, TIAS 8165(1972).

Convention on the conservation of Antarctic marine living resources. Done at Canberra, Sept. 20, 1980. U.S. Government Printing Office. 1980.

International Convention for the Prevention of Pollution from Ships. Done at London, Nov. 2, 1973. 12 I.L.M. 1319(1973).

Protocol of 1978 relating to the International Convention for the Prevention of Pollution from Ships, 1973. I.M.C.O. Document TSPP/Conf/11, Feb. 16, 1978. 17 I.L.M. 546(1978).

MARPOL Annex V. Regulations for the Prevention of Pollution of Garbage by Ships. Entered into force December 31, 1988.

United Nations Convention on the Law of the Sea. 1982. Done at Montego Bay, Dec. 10, 1982. 21 I.L.M. 1261.

APPENDIX L: Countries that Have Ratified MARPOL and its Optional Annexes (as of February 1989)

Any country that signs onto MARPOL automatically adopts the first two Annexes (which deal with discharge of oil from ships and the transport of hazardous liquids, respectively). Annexes III, IV, and V are Optional Annexes, and a MARPOL signatory country can choose whether to sign on to them. An Optional Annex comes into force only when countries representing 50% of the world's shipping tonnage have ratified it. Once it comes into effect, an Optional Annex does not apply to any country that has not ratified it, regardless of that country's participation in MARPOL. As this chart shows, several MARPOL signatories have not yet ratified Annex V, which came into force December 31, 1988 and which regulates the discharge of garbage from ships. (Annexes III and IV do not yet have the necessary 50% shipping tonnage support. They would regulate the transport of hazardous materials in packaged form (III) and the discharge of sewage from ships (IV).

State	Annexes I & II	Annex III	Annex IV	Annex V
Algeria	X	X	X	X
Antigua and Barbuda	X	X	X	X
Australia	X			
Austria	X	X	X	X
Bahamas	X			
Belgium	X	X		X
Brazil	X			
Brunel Darussalam	X			
Bulgaria	X			
Burma	X			
China	X			X
Colombia	X	X	X	X
Cote d'Ivoire	X	X	X	X
Czechoslovakia	X	X	X	X
Democractic People's Republic of Korea	X	X	X	X
Denmark	X	X	X	X
Egypt	X	X	X	X
Finland	X	X	X	X
France	X	X	X	X
Gabon	X	X	X	X

State	Annexes I & II	Annex III	Annex IV	Annex V
German Democratic Republic	X	X	X	X
Germany, Federal Republic of	X	X	X	X
Greece	X	X	X	X
Hungary	X	X	X	X
Iceland	X			
India	X			
Indonesia	X			
Israel	X			
Italy	X	X	X	X
Japan	X	X	X	X
Lebanon	X	X	X	X
Liberia	X			
Marshall Islands	X	X	X	X
Netherlands	X	X		X
Norway	X	X		X
Oman	X	X	X	X
Panama	X	X	X	X
Peru	X	X	X	X
Poland	X	X	X	X
Portugal	X	X	X	X
Republic of Korea	X			
St. Vincent and Grenadines	X	X	X	X
South Africa	X			
Spain	X			
Suriname	X	X	X	X
Sweden	X	X	X	X
Switzerland	X			
Syrian Arab Republic	X			
Tunisia	X	X	X	X
Tuvalu	X	X	X	X
USSR	X	X	X	X
United Kingdom	X	X		X
United States	X			X
Uruguay	X	X	X	X
Yugoslavia	X	X	X	X
Total Number	55	37	33	39
Percentage Tonnage*	80.92%	48.23%	40.61%	56.60%

---

*Source: Lloyd's Register of Shipping Statistical Tables, 1988.

APPENDIX M: Sample Data Card

Sample data card developed by the Center for Marine Conservation (CMC) to record types of marine debris. CMC has established a National Marine Debris Data Base to compile and analyze data from cleanups nationwide. In September 1988 more than 47,000 volunteers participated in the Data Base. The volunteers collected 977 tons of debris from 3,500 miles of

BEACH CLEANUP DATA CARD

Thank you for completing this data card. Answer the questions and return to your area coordinator or to the address at the bottom of this card. This information will be used in the Center for Environmental Education's National Marine Debris Data Base and Report to help develop solutions to stopping marine debris.

Name _____ Affiliation _____

Address _____ Occupation _____ Phone (_____)_____

City _____ State _____ Zip _____ M _____ F _____ Age: _____

Today's Date: Month _____ Day _____ Year _____ Name of Coordinator _____

Location of beach cleaned _____ Nearest city _____

How did you hear about the cleanup? _____

SAFETY TIPS
1. Do not go near any large drums.
2. Be careful with sharp objects.
3. Wear gloves.
4. Stay out of the dune areas.
5. Watch out for snakes.
6. Don't lift anything too heavy.
WE WANT YOU TO BE SAFE

Number of people working together on this data card _____ Estimated distance of beach cleaned _____ Number of bags filled _____

SOURCES OF FOREIGN DEBRIS. Please list all items that have foreign labels.

Country	Item Found
Example: *Mexico*	*plastic bottle - "Clarisol"*

STRANDED AND/OR ENTANGLED ANIMALS (Please describe type of animal and type of entangling debris. Be as specific as you can.)

_____

_____

_____

What was the most peculiar item you collected? _____

Comments _____

Thank you!

PLEASE RETURN THIS CARD TO
YOUR AREA COORDINATOR
OR MAIL IT TO:

Center for Environmental Education
1725 DeSales Street, NW
Washington, DC 20036

A Membership Organization

Center for
Marine
Conservation

♻EPA
United States
Environmental Protection
Agency

noaa

200

coastline. In all 25 states more than 60 percent of the debris items were plastic. The Center for Marine Conservation will provide data cards to all cleanup coordinators free of charge. Returned data will become part of the annual national report. To obtain copies of this data card and more information on the National Marine Debris Data Base contact CMC, 1725 DeSales Street NW, Washington, D.C. 20036.

ITEMS COLLECTED

You may find it helpful to work with a buddy as you clean the beach, one of you picking up trash and the other taking notes. An easy way to keep track of the items you find is by making tick marks. The box is for total items; see sample below.

egg cartons _HTI HTI HTI I_   Total [16]

cups _HTI HTI HTI HTI II_   Total [22]

PLASTIC	Total number of items
bags:	
trash	[ ]
salt	[ ]
other	[ ]
bottles:	
beverage, soda	[ ]
bleach, cleaner	[ ]
oil, lube	[ ]
other	[ ]
buckets	[ ]
caps, lids	[ ]
cups, spoons, forks, straws	[ ]
diapers	[ ]
disposable lighters	[ ]
fishing line	[ ]
fishing net:	
longer than 2 feet	[ ]
2 feet or shorter	[ ]
floats & lures	[ ]
hardhats	[ ]
light sticks	[ ]
milk, water gallon jugs	[ ]
pieces	[ ]
pipe thread protector	[ ]
rope:	
longer than 2 feet	[ ]
2 feet or shorter	[ ]
sheeting:	
longer than 2 feet	[ ]
2 feet or shorter	[ ]
6-pack holders	[ ]
strapping bands	[ ]
syringes	[ ]
tampon applicators	[ ]
toys	[ ]
vegetable sacks	[ ]
"write protection" rings	[ ]
other (specify)	[ ]
GLASS	
bottles:	
beverage	[ ]
food	[ ]
other (specify)	[ ]
fluorescent light tubes	[ ]
light bulbs	[ ]
pieces	[ ]
other (specify)	[ ]

STYROFOAM® (or other plastic foam)	Total number of items
buoys	[ ]
cups	[ ]
egg cartons	[ ]
fast-food containers	[ ]
meat trays	[ ]
pieces:	
larger than a baseball	[ ]
smaller than a baseball	[ ]
other (specify)	[ ]
RUBBER	
balloons	[ ]
gloves	[ ]
tires	[ ]
other (specify)	[ ]
METAL	
bottle caps	[ ]
cans:	
aerosol	[ ]
beverage	[ ]
food	[ ]
other	[ ]
crab/fish traps	[ ]
55 gallon drums	
rusty	[ ]
new	[ ]
pieces	[ ]
pull tabs	[ ]
wire	[ ]
other (specify)	[ ]
PAPER	
bags	[ ]
cardboard	[ ]
cartons	[ ]
cups	[ ]
newspaper	[ ]
pieces	[ ]
other (specify)	[ ]
WOOD (leave driftwood on the beach)	
crab/lobster traps	[ ]
crates	[ ]
pallets	[ ]
pieces	[ ]
other (specify)	[ ]
CLOTH	
clothing/pieces	[ ]

(OVER)

APPENDIX N: Bibliographic Sources of Information

There has been an increasing number of publications on the plastic debris issue. In addition to the following primary sources of reference, many public agencies and private organizations listed in the preceding appendices produce and distribute publications on their activities including information which may be relevant to the marine debris issue.

Bean M.J. 1984. United States and International Authorities Applicable to Entanglement of Marine Mammals and Other Organisms in Lost or Discarded Fishing Gear and Other Debris. Final report for the Marine Mammal Commission contract MM26299943-7. NTIS P885-160471. 65 pp. (Available from the National Technical Information Service 5285 Port Royal Road, Springfield, VA 22161.)

Center for Environmental Education. 1988. 1987 Texas Coastal Cleanup Report. Washington, DC. 105 pp. (Available from the Center for Marine Conservation, 1725 DeSales Street, NW, Washington, DC 20036.)

Cottingham, David. 1988. Persistent Marine Debris: Challenge and Response: The Federal Perspective. 41 pp. (Available from NOAA, Office of Chief Scientist, Room 6222, Washington, DC 20230.)

Interagency Task Force on Persistent Marine Debris. May 1988. 170 pp. (Available from NOAA, Office of Chief Scientist, Room 6222, Washington, DC 20230.)

O'Hara K.J. and S. Iudicello. 1987. Plastics in the Ocean: More Than a Litter Problem. Washington, DC. 128 pp. (Available from the Center for Marine Conservation, 1725 DeSales Street, NW, Washington, DC 20036.)

O'Hara K.J., N. Atkins and S. Iudicello. 1986. Marine Wildlife Entanglement in North America. Center for Marine Conservation. Washington, DC. 219 pp. (Available from the Center for Marine Conservation, 1725 DeSales Street, NW, Washington, DC 20036.)

Shomura R.S. and H.O. Yoshida (editors). Proceedings of the Workshop on the Fate and Impact of Marine Debris, 27-29 November 1984, Honolulu, Hawaii. U.S. Dep. Commer., NOAA Tech. Memo. NMFS NOAA-TM-NMFS-SWFC-54. 580 pp. (Available from the National Technical Information Service, 5285 Port Royal Road, Springfield, VA 22161.)

The Society of the Plastics Industry. 1987. Proceedings of a Symposium on Degradable Plastics, 10 June 1987. Washington, DC. The Society of the Plastics Industry, Inc. 55 pp. (Available from the Society of the Plastics Industry, Inc., 1275 K Street, NW, Suite 400, Washington, DC 20005.)

Wolfe, D.A. (editor). 1987. Plastics in the Sea. Marine Pollution Bulletin. June 1987. Volume 18. Number 6B. (Available from Pergamon Journals Inc., Maxwell House, Fairview Park, Elmsford, NY 10523.)